AERODATA INTERNATIONAL

Fighters of World War II
Volume 1

- Fw 190A
- Bf 109E
- Spitfire Mk I & II
- Hurricane Mk I
- P-51D
- P-47D

squadron/signal publications

Published 1980 by Squadron/Signal Publications, Inc.
1115 Crowley Drive
Carrollton, Texas 75006

ISBN 0-89747-109-1

Copyright © 1980 Vintage Aviation Publications Ltd.,
VAP House, Station Field Industrial Estate, Kidlington,
Oxford, England, and no part may be reproduced in any
way without the prior permission of the publishers.

FOCKE-WULF 190A
By Peter G. Cooksley

Fig. 1 *One of the first FW 190A-1s of Jagdgeschwader 26, based in Northern France, pictured on patrol over the English Channel.*
[All photographs courtesy Bruce Robertson unless otherwise credited.]

Fig. 2 *Three FW 190A-1s seen during engine run-up tests.*

Fig. 3 *The first prototype, the FW 190V1 D-OPZE, in its original form with its 1,550hp BMW 139 radial engine fitted with a ducted spinner to reduce drag.*

Fig. 4 *In 1940 the FW 190V1's ducted spinner was removed, due to engine overheating, and replaced by a tight-fitting NACA cowling. Note factory radio call sign FO+LY replacing civil registration.*

Fig. 5 *Early FW 190 prototypes and pre-production aircraft had short-span wings as seen to advantage here on the V5k, which introduced the new BMW 801C engine.*

The RAF Station at Pembrey, a few miles west of Swansea in South Wales, was all activity as technicians both Service and civil converged on the base from all over the United Kingdom to examine the German fighter there, for, on 23 June, 1942 a surprised Oberleutnant Arnim Faber had been captured by the duty officer brandishing a Very pistol when the Nazi had landed an intact Focke-Wulf FW 190A-3 after a navigational error in mistaking the Bristol Channel for the English Channel. Re-marked MP499, the machine was flown on test and the secrets of the excellent fighter which the RAF had been encountering in increasing numbers were finally laid bare.

The beginnings of these designs had been made as early as 1937 when a development contract had been placed with the Focke-Wulf company of Flughafen, the resultant prototype flying two years later under the designation FW 190V1. A second prototype, the FW 190V2, flew in the autumn of 1939 and, like the V1, it originally had a ducted spinner.

With the change of engine from the BMW 139 to the BMW 801C which, in that form or the later BMW 801D was to power the subsequent service machines, opportunity was taken to move the cockpit aft as pilots complained of the heat due to the close proximity of the engine. The poor view on the ground from this position was to cause many taxying accidents in the years to follow and the first of these was to bring about the destruction of a later prototype when it collided with a tractor.

Fig. 6 *The first pre-production batch of the FW 190 were known as A-Os, and from the eighth one onwards they featured an enlarged wing, as seen on this example, to improve manoeuvrability.*

Fig. 7 *Rear view of FW 190A-1 SB+ID showing the 190's wide-track undercarriage, which, as in the case of the Hawker Hurricane, was a distinct advantage during take-off and landing compared with the narrow-track units of the Me 109 and Supermarine Spitfire.*

Fig. 8 *Second production FW 190A-1.*

Fig. 9 *Several of the later A-O pre-production machines, including this one, were tested by II./JG26 at Maldeghem in Belgium.*

While the work was going on, construction had begun of 28 FW 190A-0s for service evaluation, the first seven retaining the short-span wings of one of the later prototypes. The majority of these, delivered at the beginning of the winter of 1940, were the subject of trials at Rechlin-Roggenthin where it was discovered that faults with the hood which might prevent escape for the pilot, cowling fasteners, and a tendency of the motor to overheat all required attention, the latter being effected by modification to the large cooling fan in front of the engine. There was also the need for some re-stressing of the aircraft. Despite these shortcomings the reports on the new fighter were largely enthusiastic, a point of particular commendation being the wide-track undercarriage which simplified ground handling on indifferent surfaces.

There followed a batch of 102 fighters for operational testing, their description being FW 190A-1, and the order was complete by the early summer of 1941. Although these were largely well received there was some disappointment with the fire power afforded by the four 7.3mm calibre MG 17s synchronised to fire through the

Fig. 10 *An early A-1, with factory call-sign markings, undergoes inspection before a test flight.*

Fig. 11 *Oblt Arnim Faber's FW 190A-3 (W Nr 313) which landed by mistake at RAF Pembrey, in South Wales, providing the Allies with valuable information on the fighter.* [via Philip J. R. Moyes.]

Fig. 12 *Faber's aircraft at the Royal Aircraft Establishment, Farnborough, in RAF camouflage and serialled MP499.* [via Philip J. R. Moyes.]

airscrew disc and the two 20mm MG FFs in the wings. This had been anticipated and in the FW 190A-2 the wing root mounted weapons were another pair of MG FFs with twin 20mm MG 151s in the outer panels.

By the time of the arrival of the FW 190A-3 in 1941 an engine change had benefited performance and the BMW 801C-2 had been replaced by the BMW 801D-2, while the Oerlikon guns in the wing roots had been moved outboard and replaced by Mausers with a higher rate of fire. Meanwhile a fighter unit in the Le Bourget area had, in May of the same year, been the first one to receive the new fighter although its baptism of fire came with an action involving Adolf Galland's JG 26 based on the Channel coast.

There followed a number of variants falling into the rebuilding category denoted by the suffix "U" from the word Umbau indicating that the conversion to a different role or structural modification had been achieved either to completed airframes or changes carried out on the production line.

A similar suffix of "R" (Rustsatz), although not introduced until the fitting of WGr 21cm rocket tubes on the A-4/R6 model, showed that the change had been brought about by means of a kit of conversion parts.

The first large-scale action involving the FW 190A was that concerned with the dash up the English Channel of the battle cruisers Scharnhorst and Gneisenau, each of 26,000 tons, together with the 10,000 ton Prinz Eugen on 12 February, 1942 under the protective umbrella of over 200 FW 190s operating with a handful of Me 109s in groups of 32 each, 16 providing the actual cover with a similar number at instant readiness; one recalls that at the time it was noted that the Nazi fighters had to lower their flaps and landing gear on occasions in order to reduce speed sufficiently to hold their sights on the obsolete British naval Fairey Swordfish torpedo machines long enough to make sure of a kill.

During the midsummer of 1942 there appeared the A-4 model which incorporated MW50, the Methanol-Wasser fuel injection system, a few such models having MG FFs in the wing roots, although this was not general; JG 2 (Richthofen) was one unit equipped with this variant. Of interest too is the fact that the A-4 was the only example of the A series to be employed in the Bv 246 "Hailstone" long-range glide-bomb experiments —the Bv 246 being a projected replacement for the Fi 103, better known as the V1 flying bomb. Although it was anticipated that the small explosive-packed glider should be flown in groups of three from bombers, for trial purposes a single such missile was launched from underneath the fighter where it was retained in position on a crutch between the undercarriage legs.

The basic version of the A-4 was also produced in a

Figs. 13, 14, 15 & 16 *More views of Faber's A-3 at Farnborough, including close-ups of the engine, nose armament, undercarriage and tail unit which will be of special value to modellers. RAF trainer-style paint scheme comprised dark earth/dark green upper surfaces and yellow under surfaces.* [via Philip J. R. Moyes.]

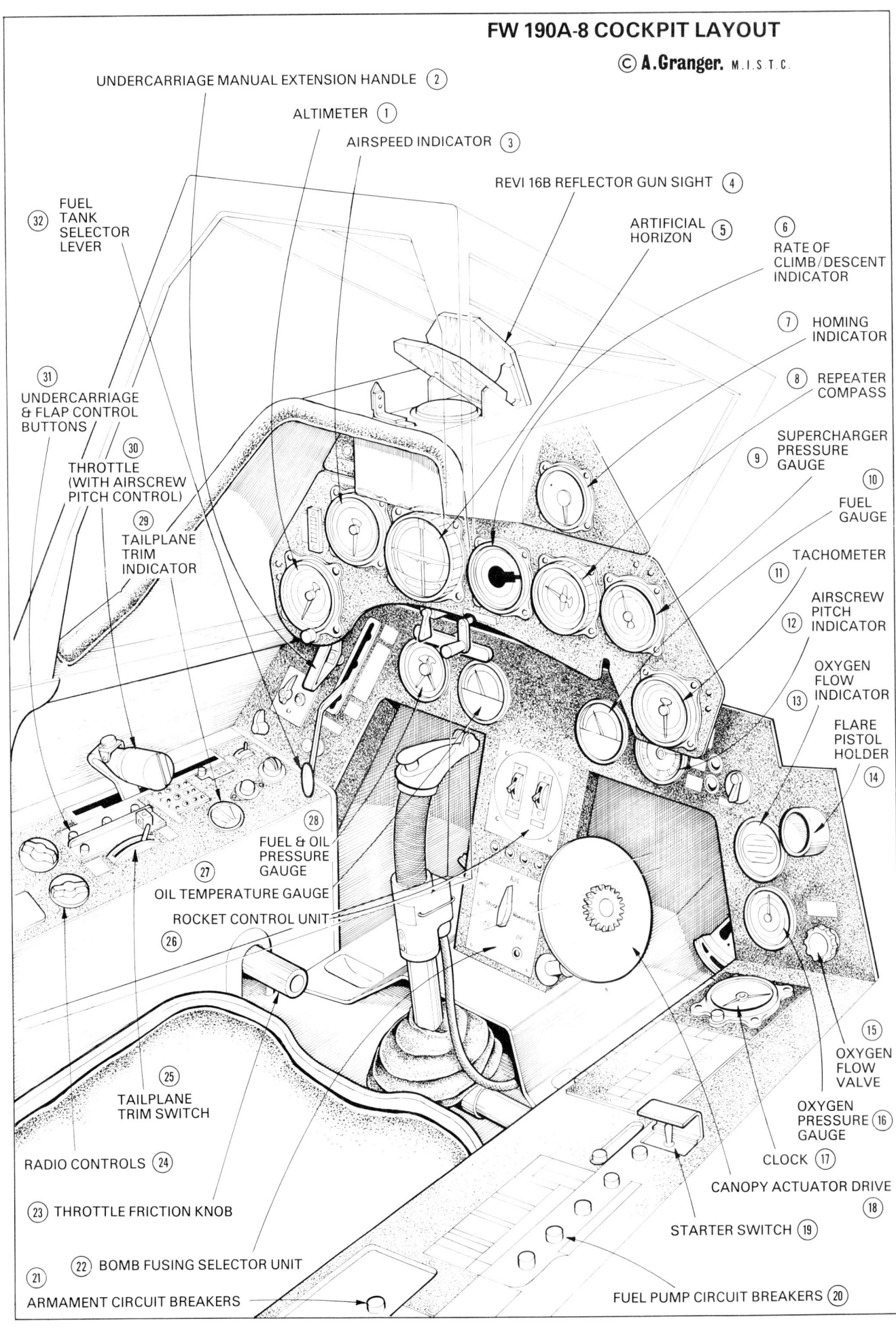

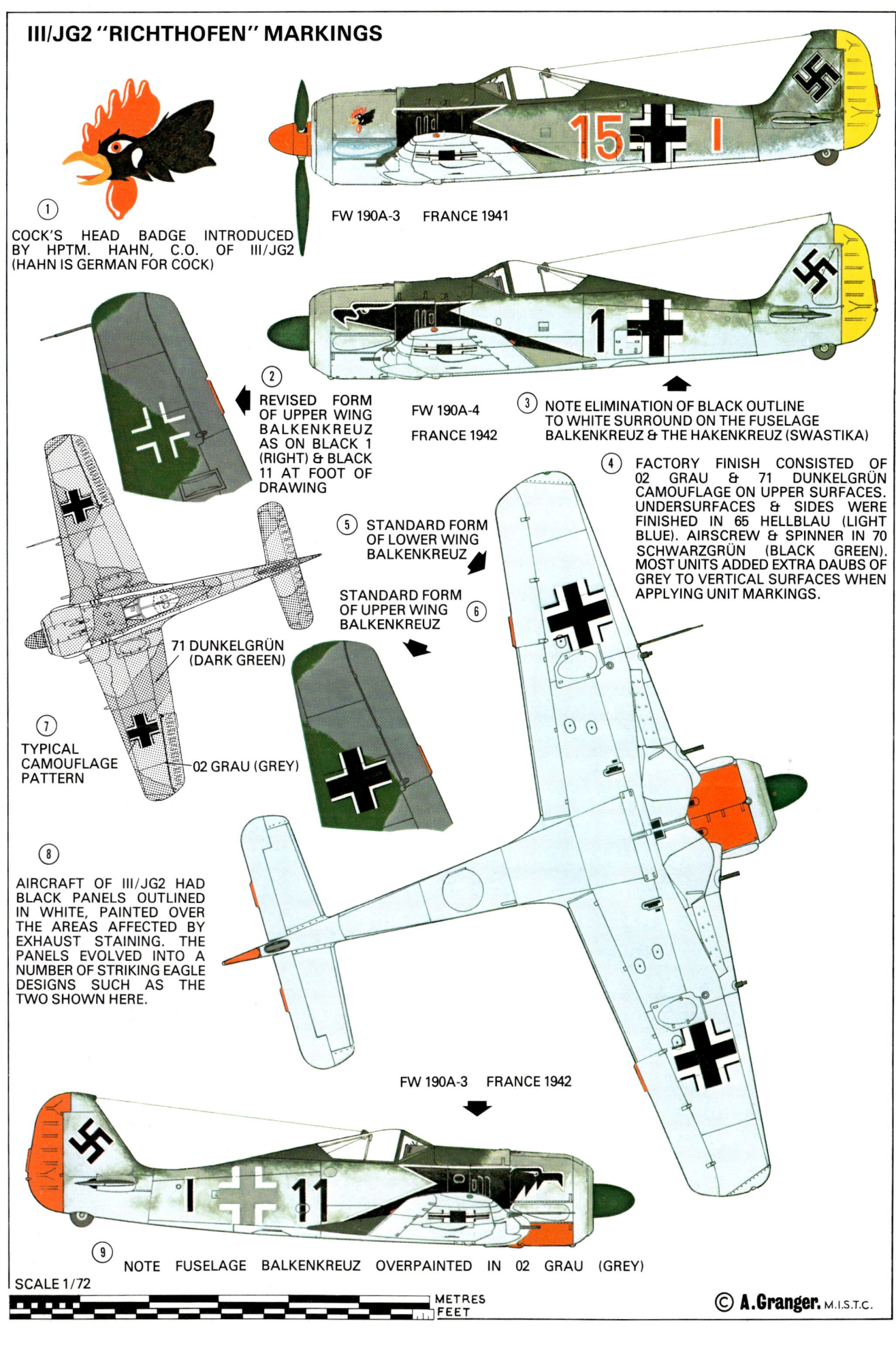

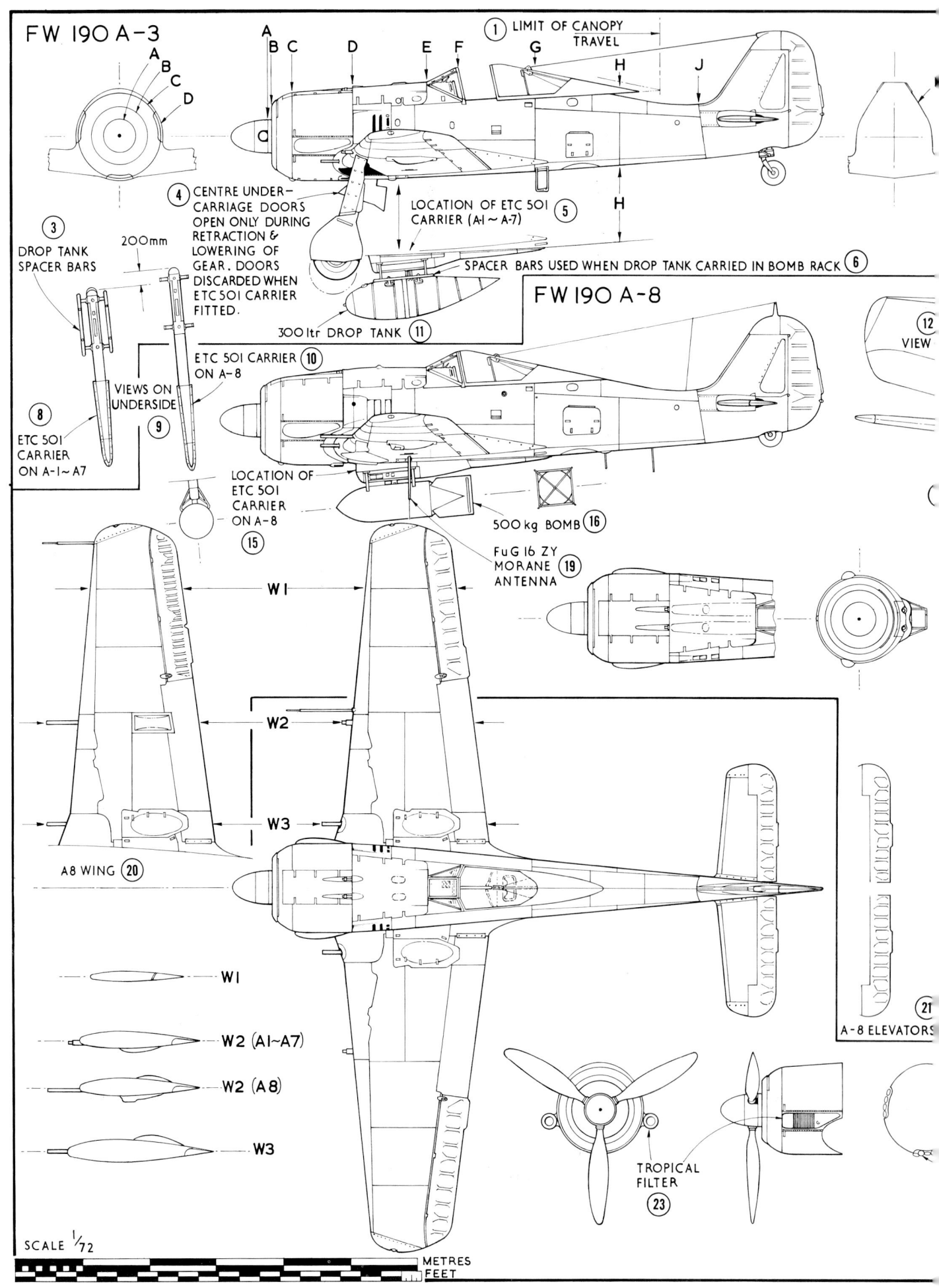

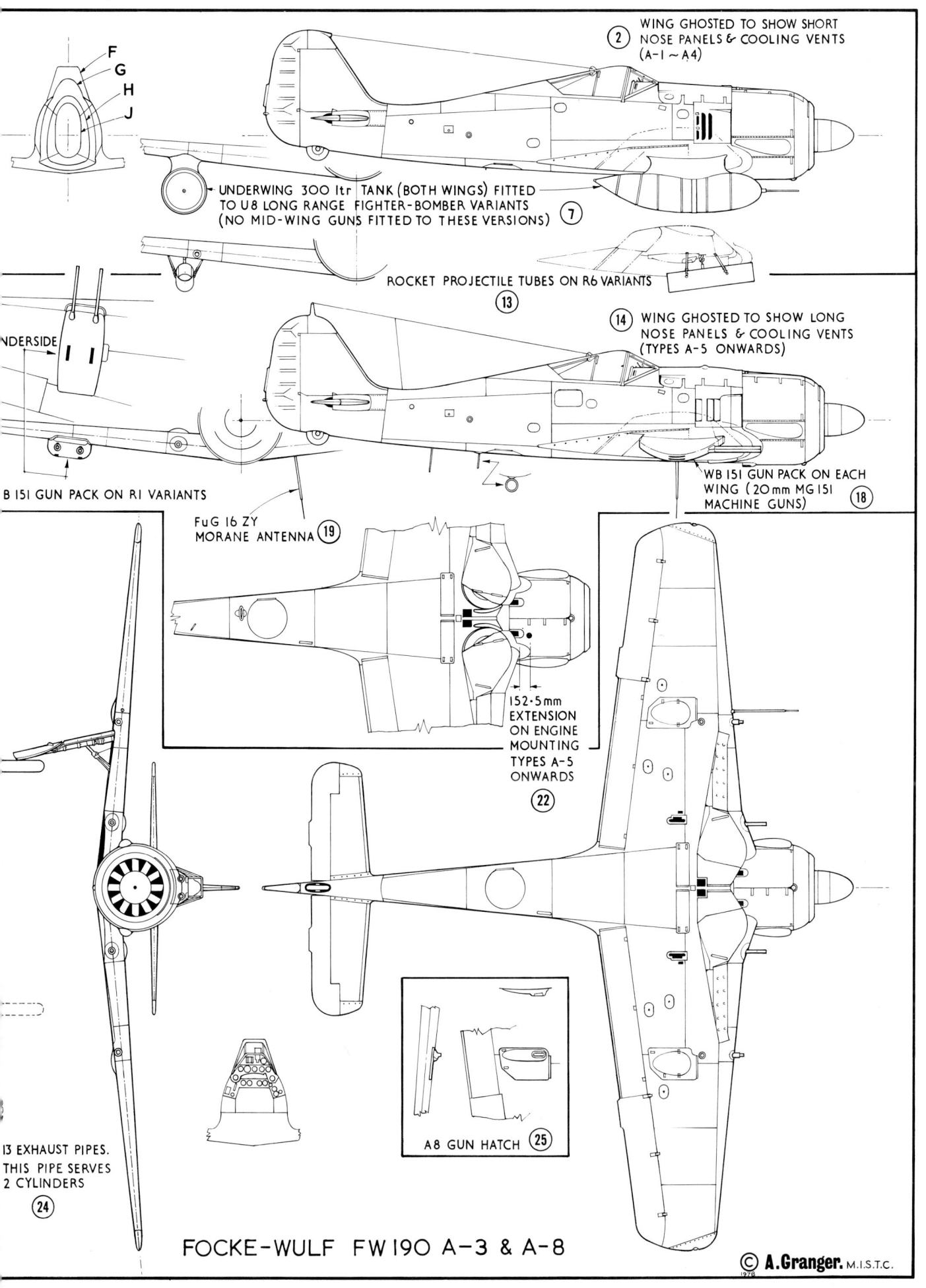

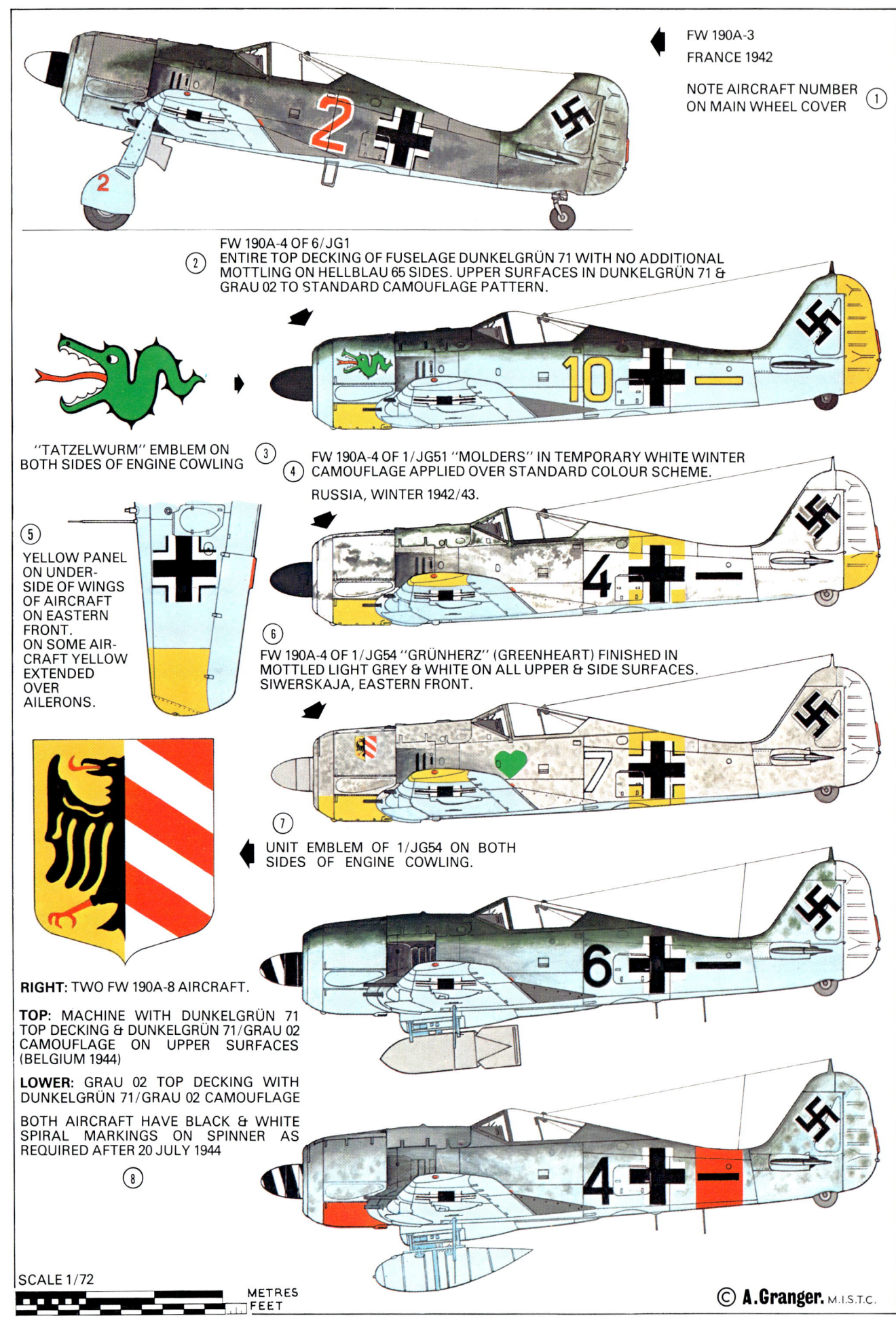

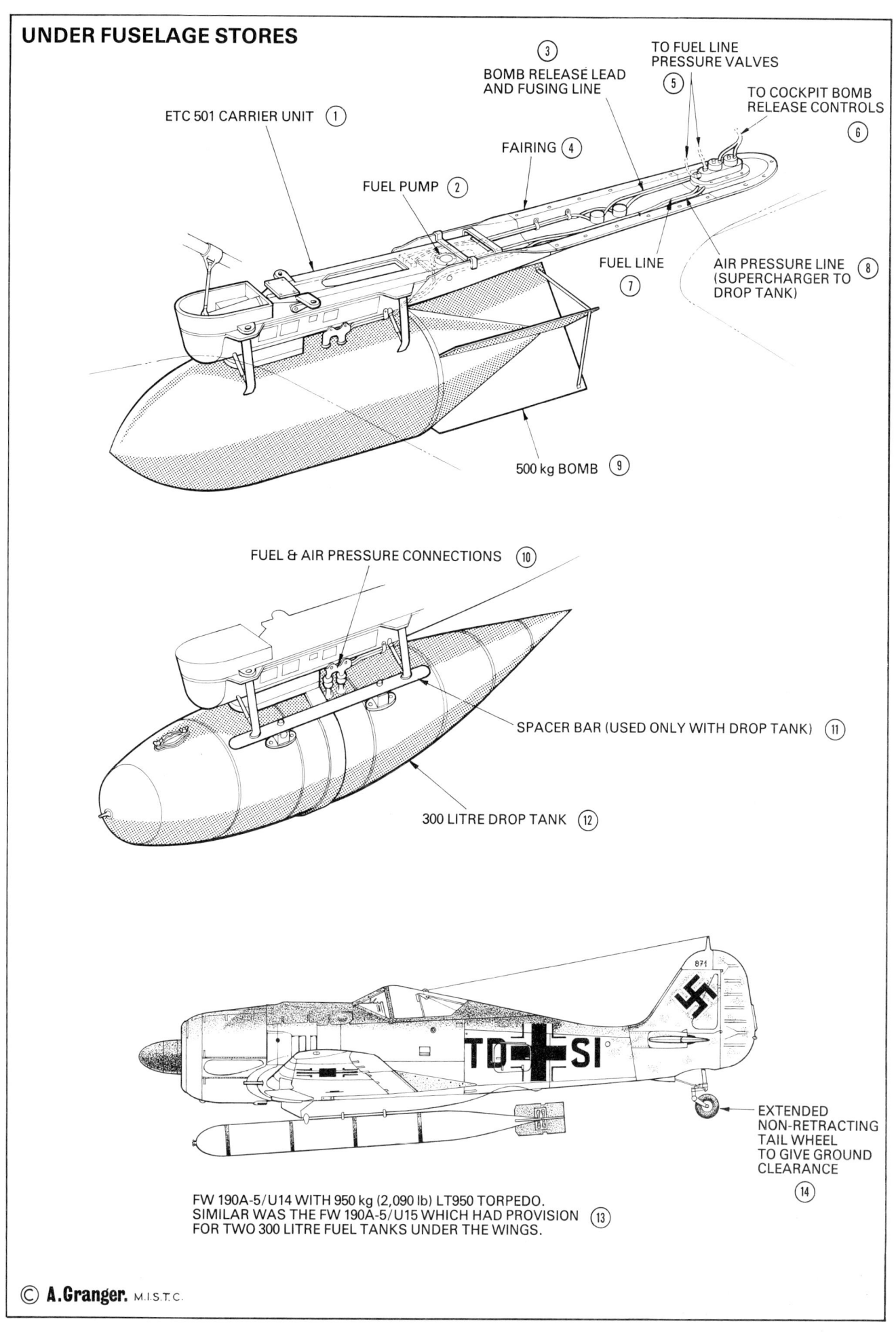

Fig. 17 *Faber's A-3 ready for take-off at Farnborough.*

tropical form which incorporated filters for operations in dust-laden climates and a rack for a 550lb (250kg) bomb. Works-rebuilt models of the basic model included the A-4/U1, a heavy fighter-bomber armed only with a pair of MG 151s—one of the best guns produced by the Mauser Works—and two ETC 501 external stores racks. The /U3 was the prototype of the later FW 190F-1 and this was followed by the /U8, a long-range fighter-bomber with provision for drop-tanks. Final sub-version of the type was the FW 109A-4 which was produced by means of the /R6 kit which allowed the carriage of WGr 21 rocket-launching tubes under the outer wing panels, while the MW50 injection was deleted.

When, during 1942, the A-5 made its appearance it was immediately remarked that the length had been increased by about six inches (152.5mm) due to a change of engine mounting but apart from this and a decreased weight it differed little from the form previously described. It did, however, provide the basis for a range of variants of great variety and in its basic form was the subject of the "Mistel" composite trials in which a fighter, which could also be of the A-8 form, was mounted above the bulk of an explosive-packed Junkers Ju 88 and released over the target by the pilot of the upper aircraft. Of even greater interest was the programme to modify several A-5 machines for use in connection with experiments involving the Gotha glider fuel tank in which a 143 gallon (650 litre) airborne tank was flown above and behind a conventional fighter by means of twin steel cables. Once the fuel in the manned aircraft was exhausted the glider could be drawn down to outriggers on the powered machine and, after its contents had been transferred, finally jettisoned; the idea was never employed operationally.

While the A-5 was a particularly rich source of prototypes for future development—the /U10 to /U12 versions becoming the 190A-6, the 190A-8/R3 and 190F-8/R3, and 190A-8/R1 with the 190A-6/R1 respectively—especially worthy of interest is the A-5/U8 version which was modified to carry a single torpedo-bomb under the fuselage. This weapon, which carried charges in its different forms varying from over 2,000lb (907kg) of explosive to in excess of only 200lb (90.7kg) was a bulky item and due to its large diameter the lower fin had to be removed to provide the machine so modfied with sufficient ground clearance. Normally a pair of SC500 bombs, each with a warhead of 595lb (270kg), were carried under the wings of unmodified heavy fighters.

By this period the FW 190 was being used in increasing numbers on all war fronts and tests were begun with a view to its employment for maritime work. Consequently the /U14 version of the A-5 was introduced with a rack under the fuselage to take a torpedo, and to facilitate ground clearance the tailwheel was mounted on a stalky extension. This type led in turn to the A-5/U15 which incorporated the lessons learnt from the previous type, but in fact only three of the latter models were built.

The RAF had by now discovered that it had a sound adversary in the German fighter and that from some points of view it had the edge over machines opposed to it. The earliest reaction to its appearance was that the choice of radial engine, a decision that had initially caused some surprise in Germany, indicated a shortage of suitable in-line engines. Recognition manuals of the period stressed the resemblance of the Focke-Wulf to the Curtiss P-36, several of which were to be seen in British skies following the French collapse, having originally been ordered for use by the Armée de l'Air. The fact that both sides were alive to this similarity is

Fig. 18 *A FW 190A-3 in service with the Turkish Air Force.*

shown by the use of captured P-36s as decoys on airfields within range of British low-level attacks where, with the added deception of German markings, they drew their share of fire.

In the theatres of war as widely separated as Russia, Africa and Great Britain the FW 190 rapidly became the symbol of the Luftwaffe's fighter arm, and in this country the type became associated with the so-called tip-and-run raids which, despite this propaganda-chosen name, were never as casual nor as innocent as the term seemed to indicate. Only the day before this text was begun an eye-witness recalled to the author the sight of half-a-dozen Focke-Wulf 190 fighters with bombs under their bellies streaking at low level up a valley in Kent with guns blazing to destroy the barrage balloons at the head of the vale before they could be raised to operational height.

Although this fighter is always remembered as the backbone of the Nazi German Air Forces, completeness demands that mention is made of the type with other nations for, across Europe, a common sight in Turkish skies was that of 190s in formation with Spitfires since both nations supplied aircraft to the Turkish Air Force. Both types of fighter retained the camouflage system in which they were delivered, but for national markings they boasted the white-outlined red square on the wings alone while the rudder was marked with a crescent moon and a single star on an entirely red background. The only fuselage marking was a large red numeral.

Surprisingly enough some 190s also flew in French colours and it is worth noting that the type was built in French factories during the occupation of that country.

Fig. 19 *An A-4, probably of IV./SKG 10, on the German airfield at Cognac in Occupied France in 1943.*

Fig. 20 *Plan view of a captured A-4/U8, serialled PE882, on test. It landed in error at RAF West Malling, Kent, during the early hours of 17 April, 1943.*

Some machines, mostly FW 190A-8s, were pressed into service following capture but after withdrawal of German forces production went ahead at the SNCA underground factory at Cravant near Auxerre where they were termed NC 900s. The first of these, an A-5, was completed as early as 16 March, 1945 and such records as survive state that No 63 made its acceptance flight on 14 March a year later, only one other, No 64, being completed. However that may be, it is known that No 62, an A-8, stands on exhibition in the Musée de l'Air and several similar machines flew with the Normandie-Niemen GC III/5 in company with Yak-3s. These may have been captured examples since the service life of the type was a short one, mainly due to the use of engine parts for the BMW 801D which had been manufactured for the Germans by French labour who found it a comparatively easy task to over-temper components so that the motors were certain to seize after even a moderate period of running. Limited funds prevented the production of fresh parts and as no alternative power plants could be procured the NC 900 quickly faded into oblivion.

Markings of the French machines were unspectacular and consisted only of the usual Tricolore roundels in the normal six positions against the camouflage finish of the period. On the fuselage, roundels had a narrow yellow rim. The rudders were picked out in the shades of the Tricolore and it is interesting to note from a surviving example that the divisions were not equal but in strict accordance to the proportions of the French national flag, namely 90:99:111, respectively blue, white and the colour at the trailing edge red. Superimposed on these appeared, in descending order, one above the other, NC 900, Type A5 (or A8) as a single line, with the serial, eg No 35, just below the mid point.

Lesser known is that a single FW 190A-5 ran the Allied blockade during the summer of 1943 and was assembled in Japan prior to a concentrated programme of flight testing and evaluation in the following September. To facilitate this, instructional manuals were translated into Japanese and the instruments re-marked. Flown against the then advanced Nakajima Ki.84 Hayate, known to the Allies as "Frank", it proved itself superior according to the report of one of Japan's foremost military pilots of the day, Major Kuroe; and Major Katakura discovered that it was possible to land it at high speed without the aid of flaps when a wrench, carelessly left in the mechanism, prevented his using them. However, final reports were less enthusiastic, for while they still especially mention the exceptional speed at maximum power and the ease of control in a dive, the manoeuvrability is described as poor. The ultimate fate of this particularly interesting specimen is obscure; it appears to have either been dismantled or possibly wrecked early in 1945 since it is

Fig. 21 *A 190A-4 of 1./SG 54 "Grünherz" carrying a 500kg SC 500 bomb.*

Fig. 22 *An A-4/U8 long-range fighter-bomber with two 300-litre drop tanks and a SC 500 bomb.* Fig. 23 *An A-5/U8 long-range fighter-bomber with a 1000kg SB 1000 bomb, lower fin of which has been removed for ground clearance.*

on record that the cowling and motor were in the Kawasaki experimental shops at Kagamigahara in March of that year and were still there when the Allies occupied the factory in September. Its sojourn in the East was not without profit, however, for it is on record that when the inverted-Vee-engined Kawasaki Ki.61 II was being re-designed as the radial-engined Kawasaki Ki.100-1a, the BMW installation was employed as a model for the Japanese product. The few photographs which survive of the type show it marked in German camouflage with the undersurface colouring brought up almost to the cockpit lip with a wavy demarcation line on the fuselage and Japanese national markings, white-edged, in the usual six positions.

At the same time as these tests in the Far East work was going on at the Focke-Wulf plant on the production of a new lightened model which was to become known as the A-6; this had a cannon with a higher rate of fire in

Fig. 24 *A FW 190A-8/R7 Assault fighter showing the armoured canopy and additional armour plate in the cockpit area.*

Fig. 25 *A FW 190A-5/U3 found by Allied troops on a former Luftwaffe airfield in Italy in September 1943.*

the outer wing positions. The /R1 modification was a heavy fighter version with eight guns, two MG 17s and six MG 151s, which entered service on the Eastern front at the end of November, 1943. Four of these weapons were carried, two in each of a couple of gondolas under the wings.

The /R2 was the inevitable fighter-bomber version and there followed a heavy fighter project known as the /R3. On the next variant the GM 1 booster fuel tanks in wings and fuselage were introduced with rocket tubes under the wings and outboard guns deleted on the A-6/R6 of late 1943. This was closely followed by a model with 13mm guns above the fuselage and a strengthened undercarriage and several Rustsatz modification models before the second type utilised in the "Mistel" trials, the A-8 with increased internal fuel capacity, arrived. Of these the /U1 is of especial interest for it represented an attempt to produce a training two-seat version as well as providing the prototype of the later 190S-8. To add the additional seat the cockpit was extended towards the rear and a slight "pinch" incorporated in the glazing in order to provide some degree of forward vision from the rear seat. Only a very few were produced, all conversions of standard A series aircraft. An A-8 was one of two FW 190s (the other was an F3) used in the interesting tests of the "Doublerider" streamlined auxiliary fuel tank which was mounted above each wing and tapered off aft of the trailing edge.

A number of sub-forms of the basic A-8 model followed until 1944 saw a change of power plant with the adoption of the BMW 801TS/TH F, but by the time the A-10 was produced it was obvious that the development potential of the basic A-type airframe had been exhausted and subsequent work was concentrated on the FW 190V 13, the prototype of the proposed FW 190B series of high-altitude fighters.

Although the Focke-Wulf 190 was a fast, adaptable and highly manoeuverable aircraft, its position in aviation history probably lies as much with the degree of technical expertise incorporated in the design than anything else. There were, for example, no cooling gills to disturb the airflow through the cowling of which the diameter was only a little under four and a half feet (1370mm), but instead a cooling fan, geared to something approaching three times the engine revs. The 14 individual exhausts were close set under the trailing edge of the cowling and the circular radiator designed to blend into

Fig. 26 *FW 190A-4 VL+FG carrying a Blohm and Voss Bv 246 glider bomb.*

Fig. 27 *The experimental FW 190A-5/U14 torpedo fighter.*

the contours of the mounting was protected by armour plate of between 5 and 3mm thickness. At the time of the British examination of the Pembrey machine interest was excited by the shock-absorbing provision of the main undercarriage and the partially retractable tailwheel. The main legs had a 15 inch (381m) travel and the gear driven by an electric motor worked on the worm and pinion principle to run a system of cables to operate the tailwheel. The tail trim was also incorporated in the highly developed electrical system and protection for the pilot was generous, including an armoured seat and a 2½in (63.5mm) thick windscreen raked back at 63 degrees, but against these refinements must be seen the fact that the duration of the FW 190 was only in the region of two hours and the first reports on the captured machine considered the general finish to be poor.

Fig. 28 *This NC 900, French-built version of the FW 190A-8, is preserved in the Musée de l'Air in Paris.*

Fig. 29 *A-5/U12 W Nr 813, carrying four 20mm MG 151 cannon in twin underwing packs, was used for armament trials.*

THE FW 190A SERIES

Type	Remarks
FW 190A-0	Pre-production machines for evaluation.
A-1	Initial 102 for operational testing.
A-2	Wing span increased by 7½in (190.5mm).
A-3	Improved armament.
A-3/U1	Fighter bomber. MG FFs replaced by bomb racks.
A-3/U3	Increased weight and speed.
A-3/U4	Reconnaissance fighter.
A-3/U7	Fighter bomber with MG FFs. replaced.
A-4	MW50 fuel injection system.
A-4/Trop.	With filters and rack for 550lb (250kg) bomb.
A-4/U1	Twin MG 151s plus a pair of bomb racks.
A-4/U3	Ground support fighter, became FW 190F-1.
A-4/U8	Long range fighter bomber with drop-tanks.
A-4/R6	No MW50. Fitted with rocket tubes.
A-5	Length increased by 6in (152.5mm).
A-5/U2	Night fighter.
A-5/U3	Ground support.
A-5/U4	Reconnaissance fighter.
A-5/U8	Long range fighter bomber.
A-5/U9	Twin MG 151s plus Twin MG 131s.
A-5/U10	Became FW 190A-6.
A-5/U11	Heavy fighter, became FW 190A-8/R3 and F-8/R3.
A-5/U12	Heavy fighter, became FW 190A-8/R1 and A-6/R1.
A-5/U13	Long range fighter bomber, became FW 190G-3.
A-5/U14	Experimental torpedo fighter.
A-5/U15	Special torpedo fighter, three built.
A-5/U16	Special bomber destroyer with 30mm MK 108 guns.
A-5/U17	Became FW 190F-3.
A-5/R6	Fitted with WGr 21 cm rocket tubes.
A-6	Airframe lightened.
A-6/R1	Heavy fighter with eight guns.
A-6/R2	Fighter bomber version of A-6.
A-6/R3	Heavy fighter project.
A-6/R4	GM 1 fitted.
A-6/R6	Fitted with WGr 21 cm rocket tubes.
A-7	Modified armament and strengthened u/c.
A-7/R1	Heavy fighter.
A-7/R2	30mm Mk 108 guns outboard.
A-7/R3	Provision for single drop-tank.
A-7/R6	Increased gross weight.
A-8	Increased internal tankage.
A-8/U1	Trainer, became FW 190S-8.
A-8/U11	Fighter bomber.
A-8/R1	Twin MG 151s in underwing packs.
A-8/R2	GM 1 version with increased gross weight.
A-8/R3	Ground support. Believed project only.
A-8/R7	Assault fighter.
A-8/R8	Assault fighter with modified armament.
A-8/T11	All-weather fighter.
A-8/R12	Projected version with modified armament.
A-9	Engine change.
A-9/R8	Assault fighter project.
A-9/R11	All-weather fighter with turbo-supercharger.
A-9/R12	All-weather fighter project.
A-10	Long range escort fighter.

SPECIFICATION — FW 190A-3 and A-8

Dimensions: Span 34ft 5⅜in (10500mm); length (A-1, A-2, A-3, A-4) 28ft 10⅜in (8798mm), (A-5, A-6, A-7, A-8, A-9) 29ft 4⅜in (8950mm); wing area 196.98sq ft (18.3 sq m).

Powerplant: BMW 801D-2 14 cylinder twin-row radial producing 1,580hp at 2,700rpm at take off and 1,760hp at 3,000rmp at 18,000ft (5500m).

Performance: Max speed (with over-ride boost from methanol-water 50 per cent injection (MW50) which increased engine power for a continuous period of ten minutes below rated altitude on A-4 type) 395mph at 17,000ft (636km/h at 5200m), 326mph at 4,500ft (525km/h at 1400m), 390mph at 20,000ft (628km/h at 6100m); A-8 405mph at 20,500ft (652km/h at 6250m), 355mph (571km/h) at sea level.

Rate of climb: to 16,500ft (5030m) 4.75min, to 18,000ft (5486m) 6.25min.

Service ceiling: 37,000ft (11,300m) with normal fuel load.

Range: 500 miles (800km) on internal fuel.

Weights: A-3 loaded weight 8,378lb (3800kg); A-8 loaded weight 9,452lb (4287kg), with additional 25.3 imp gall (115 litre) fuselage fuel tank fitted 9,700lb (4400kg), A-8 max permissible take off weight (using drop tanks, etc) 10,803lb (4900kg).

Tankage: 115.5 imp gall (525 litre). Additional 25.3 imp gall (115 litre) fuselage tank could be fitted to A-8; A-3 10 imp gall (45 litre) oil; A-8 12 imp gall (55 litre) oil.

Armament: A-3—twin synchronised 7.9mm Rheinmetall-Borsig MG17 machine-guns above the cowling; twin synchronised 20mm Mauser MG151 cannon in the wing roots and twin 20mm Oerlikon MGFF cannon in the wing panels. Total fire power 610lb/min (276.7kg/min. Rate of fire: MG17 machine-guns, 600 rounds per minute; MG151 cannon, 700 rounds per minute; MGFF cannon, 450 rounds per minute. Ammunition: MG17 1,000 rounds per gun; MG151 230 rounds per gun; MGFF 60 rounds per gun.

A-8—two MG131 13mm machine-guns above the cowling (with 475 rounds per gun), two MG151 20mm cannon in the wing roots (with 250 rounds per gun), two MG151 20mm cannon in the wing panels with 140 rounds per gun; A-8/R1 four MG151/20Es with 125 rounds per gun (two beneath each wing) replacing the outboard MG151s; A-8/R2 two MK108 30mm cannon with 55 rounds per gun replacing the outboard MG151s; A-8/R3 two MK103 30mm cannon with 35 rounds per gun (beneath the wings) replacing the outboard MG151s.

SUPERMARINE SPITFIRE I & II
By Philip J. R. Moyes

Fig. 1   *No 19 Squadron's Spitfire Is on parade at Duxford, Cambs., for the benefit of the Press, 4 May 1939.*

Fig. 2 *The prototype Spitfire, K5054, in flight in September 1937 after having been converted to production standard and camouflaged.*

Fig. 3 *In this view the prototype is seen in the form in which it first flew.*

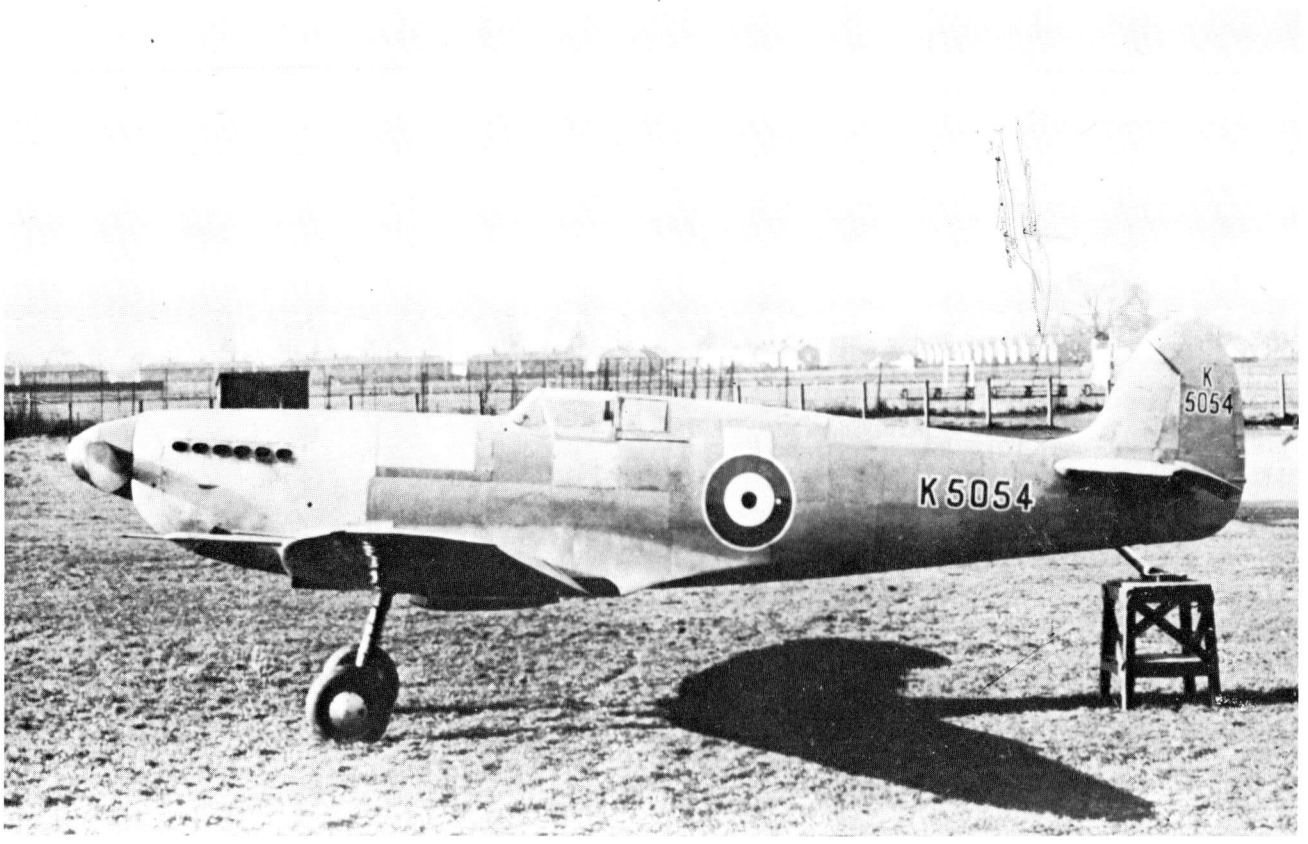

Together with the Hawker Hurricane, the Spitfire is remembered above all else for winning the Battle of Britain. Both aircraft were truly great fighters, but the Spitfire, having all the grace and beauty—and in such a generous measure, thanks largely to its classical elliptical wing—captured the public's imagination to a greater degree than any other aeroplane has or, perhaps, ever will. It became a symbol of victory wherever it flew in World War 2, and happily several specimens are still able to take to the air today to remind us of the type's glorious achievements.

The Spitfire was the brainchild of Reginald J. Mitchell, who, from 1925 onwards, evolved a series of high-speed Supermarine seaplanes of advanced design for the international Schneider Trophy Contests culminating in the S.6B, one of which on 13 September, 1931 won the Schneider Trophy outright for Great Britain at an average speed of 304.08mph (547,29km/h). Soon afterwards, when a special sprint engine had been installed, this particular S.6B (S1595) raised the world's speed record to 407.5mph (655,79km/h) and today it can be seen, preserved for posterity, in the Science Museum in London.

With these successes behind them, Mitchell and his team turned their attention to fighters, beginning with the Supermarine Type 224 in an attempt to meet the Air Ministry specification F.7/30. This was a gull-winged monoplane day and night fighter with a fixed "trousered" undercarriage and a Rolls Royce Goshawk engine; but through no fault of Mitchell and his team it fell short of the required performance. Mitchell was dissatisfied with it even before it flew (which was in February 1934) and, with the help of his colleagues, he began to design a new single-seat fighter, aiming far ahead of those hitherto conceived in Air Ministry specifications. At first this project was a company-funded private venture, but gradually Air Ministry interest was attracted and, by a continuous process of development and refinement, there eventually emerged, early in 1936, the Supermarine Type 300 Spitfire prototype, whose construction was fully covered by an Air Ministry contract.

Powered by a 990hp Rolls-Royce Merlin C engine and carrying an RAF serial number (K5054) and markings, but otherwise mostly unpainted, the new fighter made its first flight on 5 March, 1936 from Eastleigh Airport, in the hands of J. "Mutt" Summers, chief test pilot of Vickers Aviation Ltd. (Vickers had acquired Supermarine in 1928.) By experimenting with different designs of wooden fixed-pitch propellers, and also by some other design refinements, Summers and two other test pilots, working as a team, managed to increase K5054's maximum speed from 336 to 349mph (541-562km/h) before sending it, in April 1936, for its initial RAF trials at Martlesham Heath. On its return Supermarine made several small alterations which had been suggested by the RAF test pilots, including one to the rudder horn balance, and they also painted the aeroplane in pale grey-blue high gloss enamel. In its revised form it was flown on 11 May, 1936 by Jeffrey Quill, one of the team already mentioned. An ex-RAF pilot, Quill became Supermarine's senior test pilot in 1938 and

Fig. 4 *Take-off shot of K5054 when fully painted and fitted with hinged wheel fairings and redesigned rudder with smaller horn balance. Pilot is Jeffrey Quill.*

Fig. 5 *P9450, the 601st Spitfire I, on acceptance trials in the spring of 1940, flown by Supermarine test pilot George Pickering.*

was to be closely associated with the testing of successive Spitfire variants for most of World War 2. He is on record as saying that, contrary to oft-expressed reports, the Spitfire was *not* a trouble-free design and *did* suffer its share of teething troubles, like most new and untried aeroplanes.

In May 1936, the new fighter was officially given the name Spitfire—a name which, incidentally, had tentatively been given by the chairman of Supermarine to the unsuccessful Type 224—and on 3 June the Air Ministry placed an order for 310 Spitfires. This presented Supermarine with a tremendous task, for it

Fig. 6 *Spitfire I K9849, from the first production contract, on test before the war.*

Fig. 7 *Spitfire Is being refuelled at a Scottish base of Fighter Command in March 1940.*

had never before received such a large order and, furthermore, its production facilities on the banks of the River Itchen, at Southampton, were geared only to the production of handfuls of flying-boats. In the event, Supermarine decided to undertake the construction of Spitfire fuselages only, together with the assembly and testing of the aircraft, and to "farm out" the rest of the construction to numerous sub-contractors.

The emergence of the first production machines was considerably delayed by late deliveries of some of the sub-contracted components and various technical problems, and not until 4 August, 1938 did the Spitfire begin to enter squadron service, with the delivery of the third production aircraft to No 19 Squadron at Duxford. Not least of the production problems were those presented by the elliptical wings, and eventually Supermarine found it necessary to build these itself. As the snags were overcome, so did the flow of Spitfires from Supermarine's final assembly lines at Eastleigh Airport improve, and by the outbreak of World War 2 over 300 had been built. Orders had already risen to 2,160 Spitfires, including 1,000 due to be built at a new shadow factory at Castle Bromwich, run by the Nuffield Organisation. By that time, too, nine RAF squadrons were fully armed with Spitfires (Nos 19, 41, 54, 65, 66, 72, 74, 602 and 611) and two others (Nos 603 and 609) were in the process of re-arming.

The early production Spitfires were similar in outline to the prototype as it originally existed (it was itself modified up to Mk I standard in September 1937) but there were many detail differences. For instance, the semi-circular, hinged wheel fairings were removed, the flush exhausts replaced by triple ejector exhausts and the tailskid replaced by a castoring tailwheel. A 1,030hp Merlin II was installed in the first 174 aircraft and the two-blade fixed pitch propeller was replaced, from the 78th machine onwards, by a three-blade two-speed de Havilland model. Early in 1939 a domed cockpit hood was introduced, and among subsequent modifications were improved protection for the pilot and the introduction of a redesigned aerial mast.

From the 175th Spitfire I onwards the Merlin III engine was installed. Although of similar power to the Merlin II, this had a standardised shaft for de Havilland or Rotol propellers. Between late June and early August 1940, all de Havilland propellers on Spitfire Is were converted to constant speed units and these significantly improved the aircraft's climb rate and ceiling—and in turn did much to help smash the Luftwaffe's massive attacks during the Battle of Britain.

Altogether, 1,583 Spitfire Is were built—mostly by Supermarine although 50 were built by Westland, which latter company started deliveries in July 1941.

Before the outbreak of World War 2 several foreign governments planned to buy Spitfires, and export versions of the Mk I were designed for Estonia, Greece, Portugal, and Turkey. Foreign Office sanction to export these aircraft was withdrawn when war became imminent and only three were exported; they all went to Turkey, one of them having originally been consigned to Polland and diverted to Turkey after the Polish collapse.

Fig. 8 *Spitfire I fuselages and tail units being "married-up" in Supermarine's Itchen works at Southampton early in 1939. Aircraft in background have bulged cockpit hoods fitted.*

Fig. 9 *Another close-up shot taken pre-war in one of the erecting shops at the Itchen works.*

First kills achieved by Spitfires were nothing less than two friendly Hurricanes which, on the morning of 6 September, 1939 were mistaken for enemy aircraft and shot down during an unfortunate episode which is recorded in the annals of the RAF as the "Battle of Barking Creek". Casualties resulting from misidentification of friendly aircraft were by no means uncommon during the war, as was seen, for example, on the last day of August 1940—in the middle of the Battle of Britain—when a Spitfire I of No 54 Squadron was shot down over Hildenborough, Kent, by a Hurricane.

Meanwhile, on 16 October, 1939 three Spitfires of No 603 Squadron jointly became the first UK-based aircraft to shoot down an enemy aircraft in World War 2. Operating from Scotland, they bagged a Junkers Ju 88 which fell into the sea off Port Seton, but for some unexplained reason the victory was officially credited to only the leading Spitfire of the section involved—the machine flown by Sqn Ldr E. E. Stevens.

Less than a fortnight later, on 28 October, Spitfires of Nos 602 and 603 Squadrons spotted a lone Heinkel He 111 over the Firth of Forth and attacked it in turn. Two of the bomber's crew were killed and the pilot wounded, leaving only the observer unhurt, but somehow the pilot managed to force-land his machine in the heather close to the village of Humbie, near Dalkeith; this Heinkel, from Kampfgeschwader 26, was the first enemy aircraft to be brought down on British soil since World War 1. Less than a month later, on 20 November, Spitfires gained a further distinction—the first success

Fig. 10 *Fighters up. Two vics of Spitfire Is of No 19 Squadron on patrol early in 1939.*

Fig. 11 *Fine profile of P9450, also seen in Fig. 5, on test. This Spit was part of the first production order, which was placed with Supermarine on 20 April 1939 and fulfilled in 1940.*

Fig. 12 *Head-on view of a Spitfire I with a three-bladed DH two-speed propeller, showing to good effect the Spit's low frontal area, including the extremely thin wing. Narrow-track undercarriage was one of the Spit's least desirable features as it caused handling problems, especially during taxying when the pilot's difficulties were further increased by the long nose which completely blocked the forward view.*

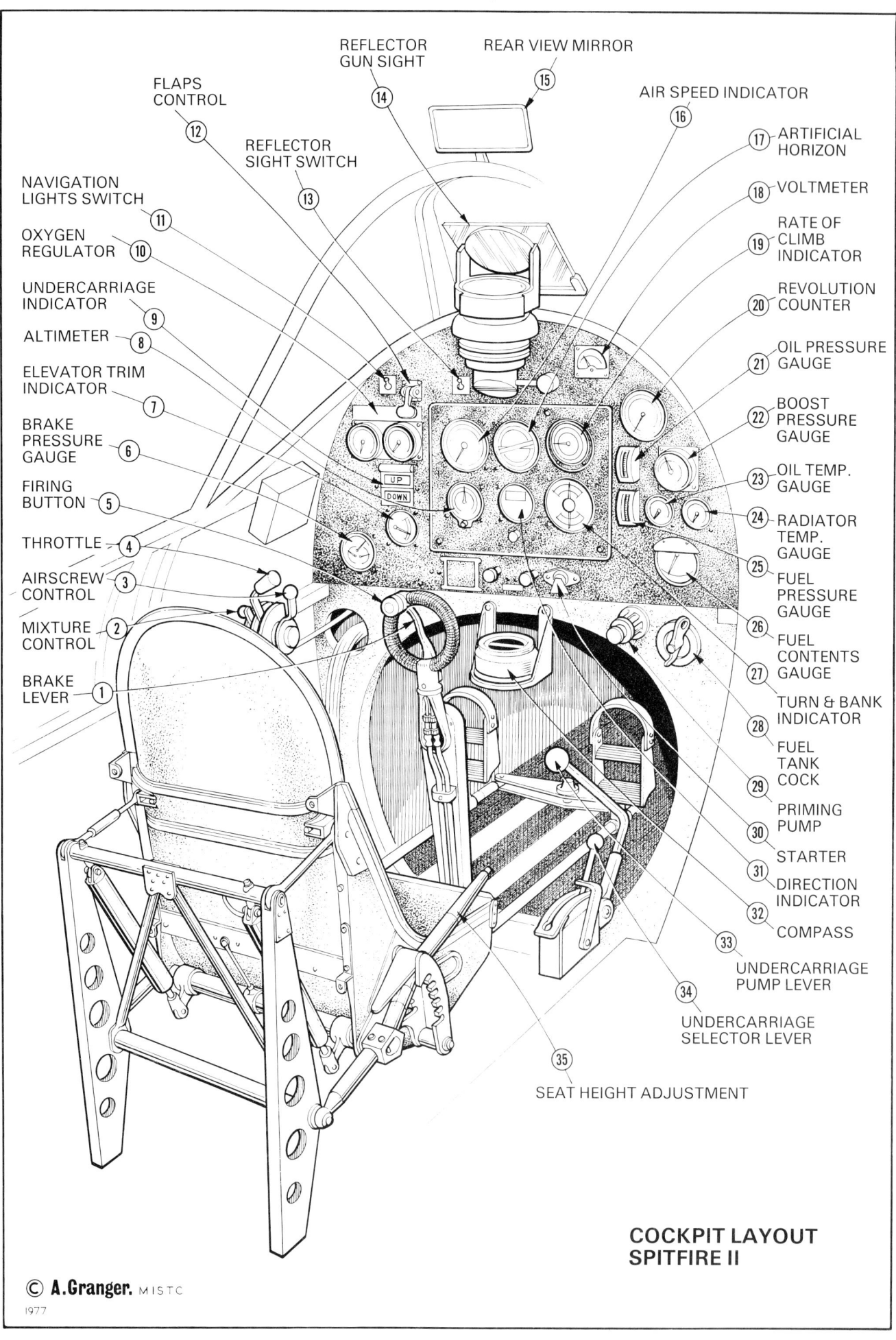

SPITFIRE I
NO 19 SQUADRON
DUXFORD
OCTOBER 1940

SCALE 1/72

ROY MILLS

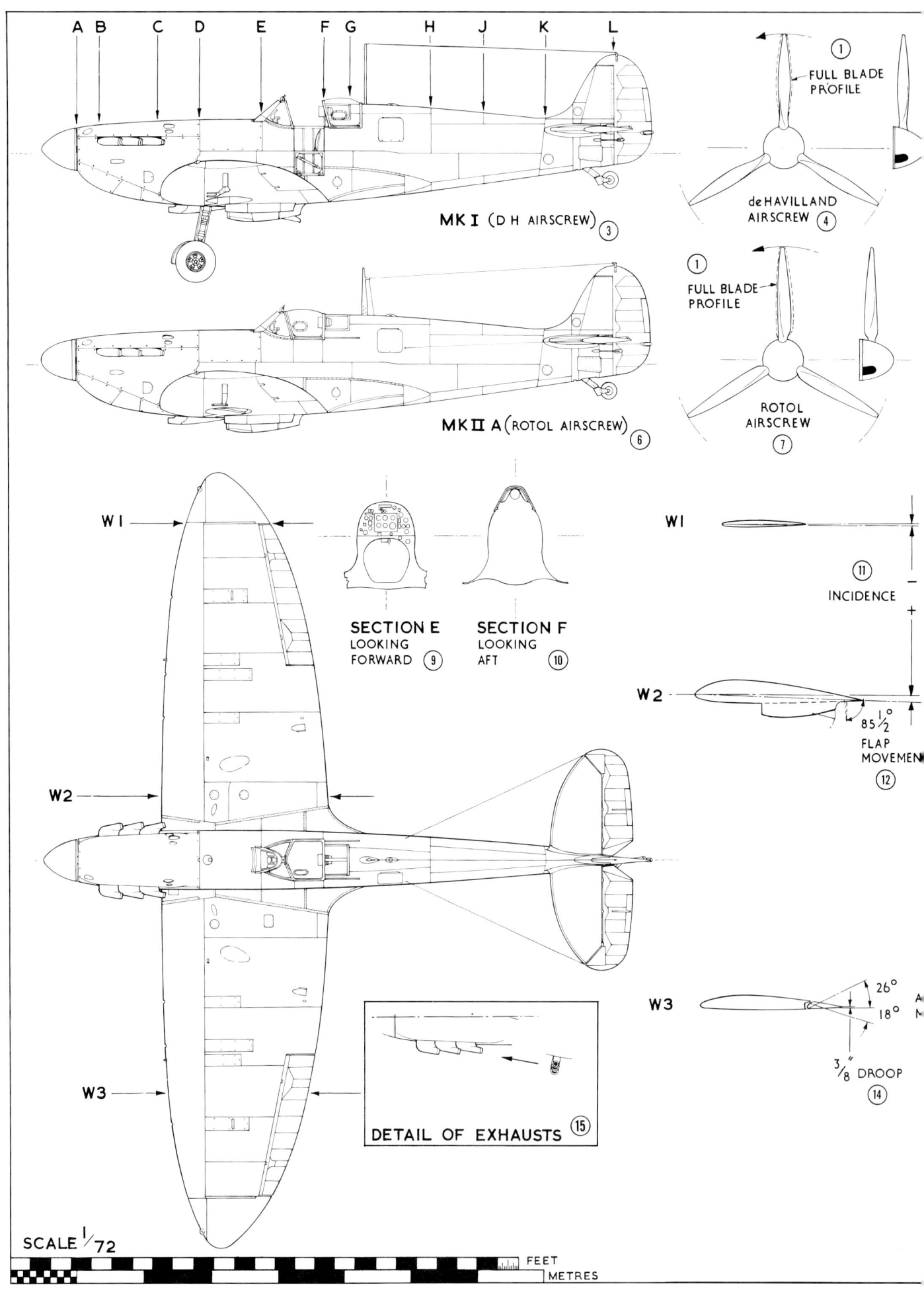

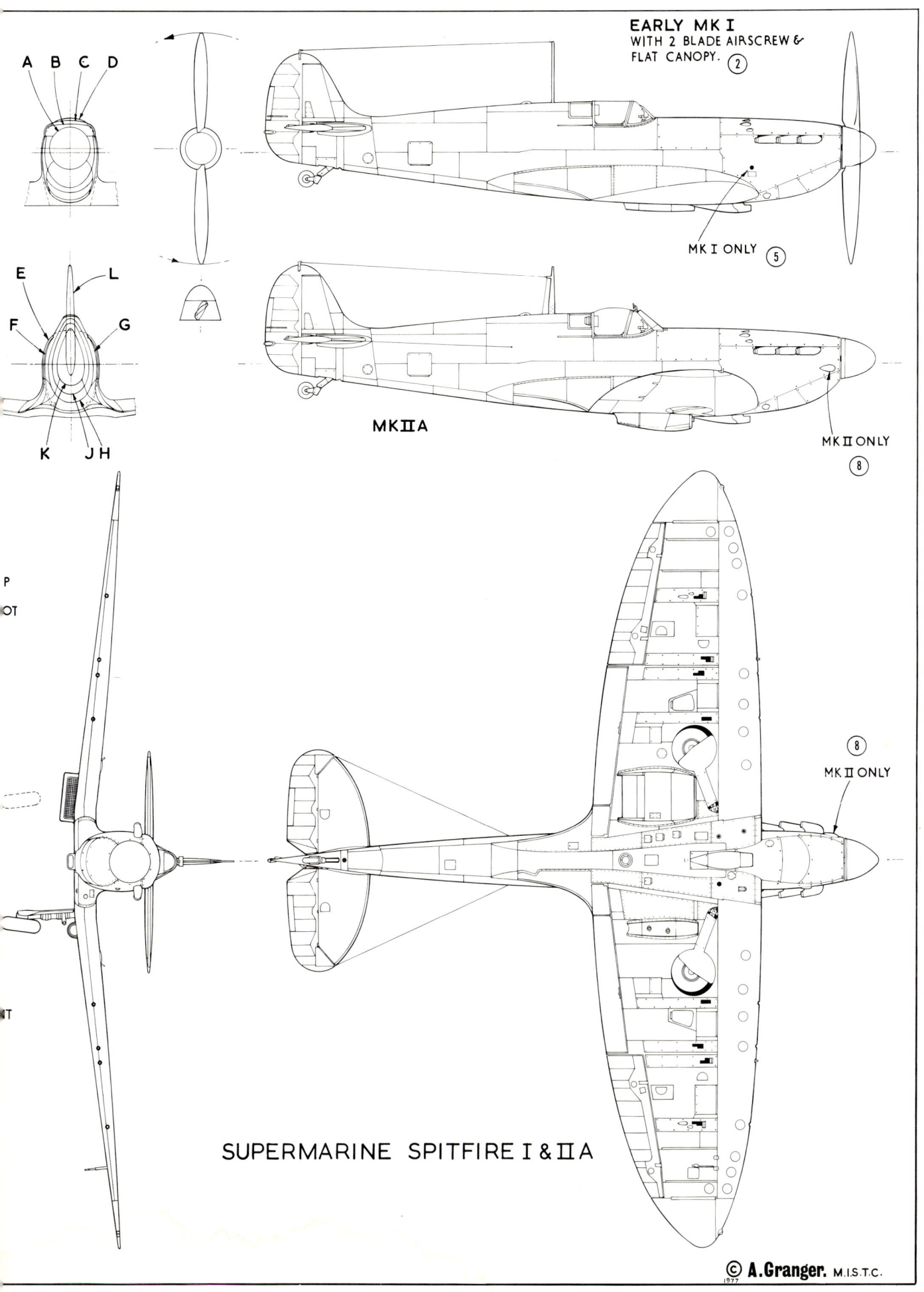

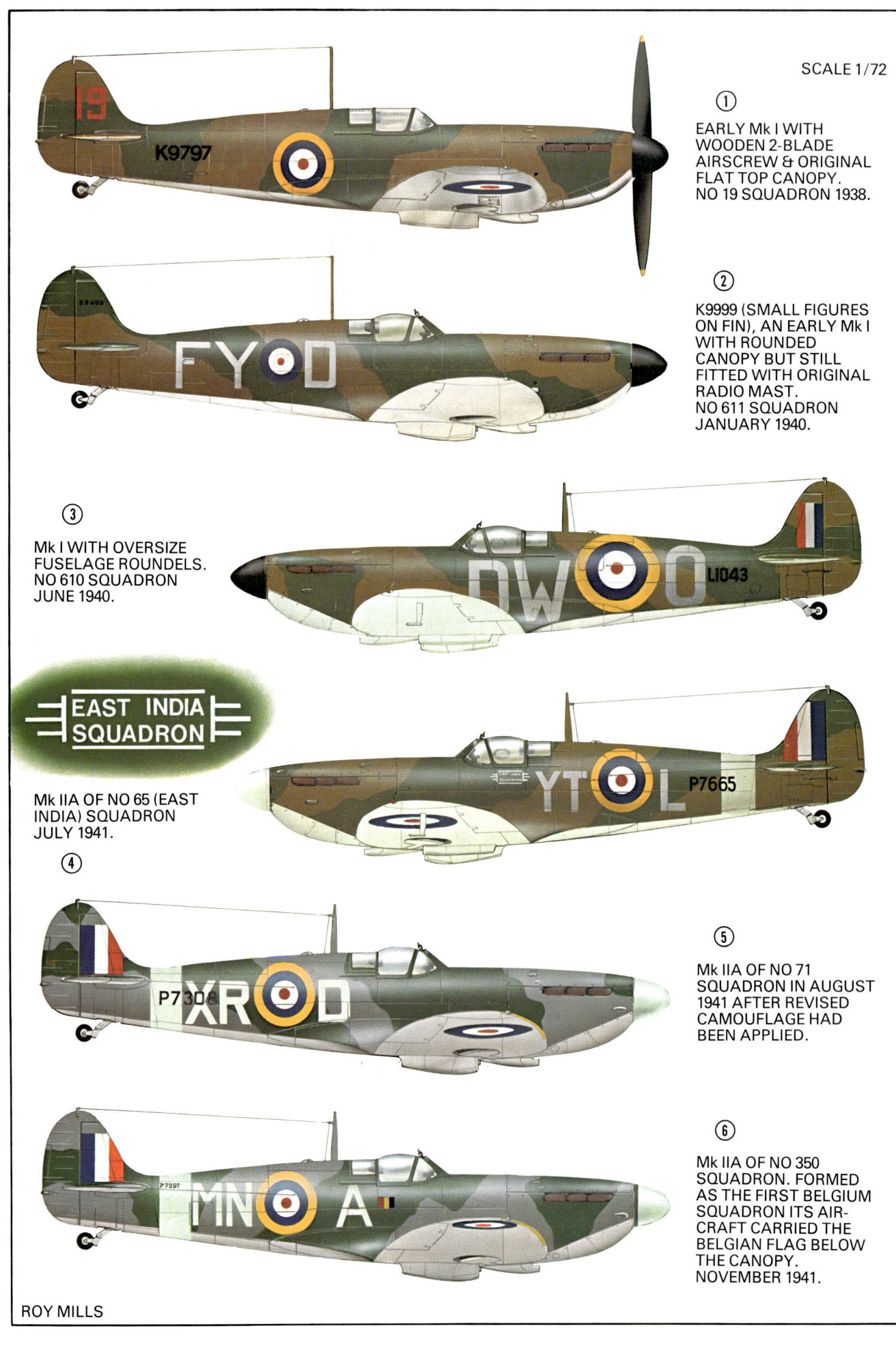

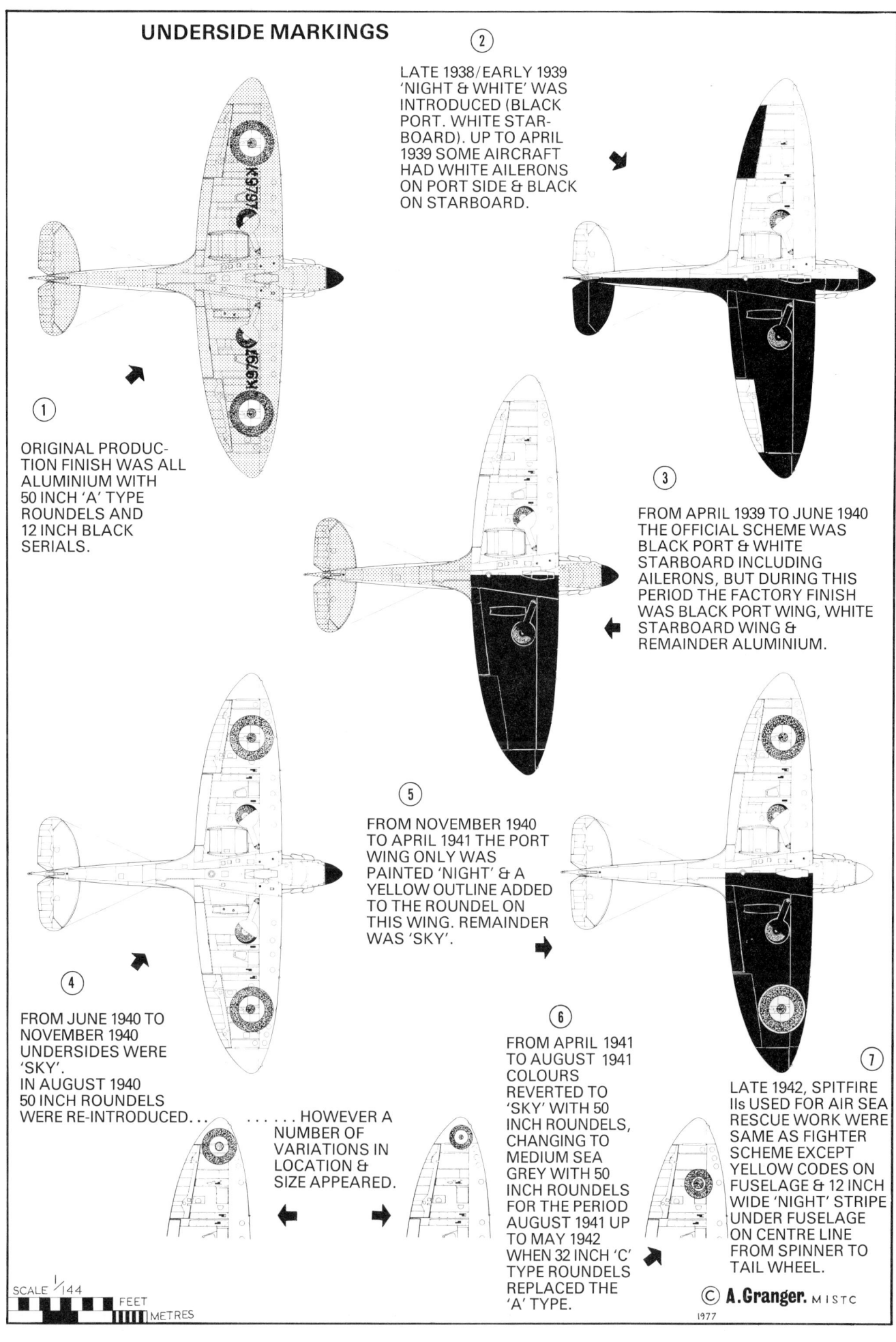

Figs. 13 & 14 *Two views showing the very neat, compact installation of the Rolls-Royce Merlin engine in the Spitfire.*

Fig. 15 *Spitfire IIA LO-J of No 602 Squadron, probably at Drem in Scotland, has its guns re-loaded after a sortie in March 1940.*

from England—when three aircraft of No 74 Squadron, operating from Essex, shot down an He 111 some 15 miles out to sea off Southend. At first they only claimed it as a probable as it was last seen entering cloud, but the discovery next day of two German airmen in a dinghy confirmed beyond doubt that the raider had been destroyed.

When the Battle of Britain began there were 19 Spitfire squadrons on the strength of RAF Fighter Command and the part the "Spits" played in that epic struggle, together with that of the more numerous "Hurries", is now a matter of history. There were still only 19 squadrons of Spitfires on strength at the end of September 1940 when the enemy's heavy daylight bomber attacks virtually ceased and the pattern of combat switched instead to fights at high altitude between fighter and fighter-bomber formations. The Spitfire, unlike its fighting partner, was able to hold its own with the Messerschmitt Me 109E in almost every respect. However, as a result of the enemy's moves to provide his aircraft with more armour plate and leak-proof, fire-resisting fuel tanks, it was becoming increasingly clear that the Spitfire's armament of eight .303in (7,7mm) Browning machine-guns was not enough; it needed the greater hitting power of cannon armament, a feature which the Me 109 already possessed. This had in fact already been forseen, for a trial installation of two 20mm Hispano cannon had been made in the wings of

Fig. 16 *A Spitfire I of No 19 Squadron pictured in 1940 when the unit had the wartime code letters QV.*

Fig. 17 *A Spit I of No 611 Squadron tests its guns at the stop butts at RAF Digby, Lincs, in January 1940.*

an early Spitfire I as early as June 1939. Subsequently, a small batch of twin-Hispano-armed Spitfire Is was put into production and some of these were tested operationally during the Battle of Britain by No 19 Squadron. When the cannon worked the results were spectacular; sad to relate, however, the Hispanos were so plagued with stoppages that the aircraft had to be withdrawn. The actual cannon were relatively blameless, the stoppages being mainly due to feed problems with their drum-type magazines and the jamming of empty shell cases during wing flexing in tight turns. In an effort to improve matters, Supermarine produced a trial installation on a Spitfire I of wings containing two cannon and four Browning machine-guns with a draw feed to the cannon. Several of the cannon-armed Spitfires that had been used by No 19 Squadron were then modified to a similar standard and issued, in the late autumn of 1940, to No 92 Squadron for operational trials; but the extra weight handicapped their performance, and these Spits too were soon withdrawn.

With the adoption of the twin cannon and four machine-gun armament for the Spitfire I—albeit temporarily—two of the three basic Spitfire wing designations became established. The original wing mounting eight Browning guns became known as the A wing, while the wing with two cannon and four machine-guns became the B wing; the two distinct types of Spitfire I thus became Mk IA and Mk IB respectively.

Naturally, Supermarine's factories at Southampton received the attention of the Luftwaffe, and in a determined bombing attack by a large force of He 111s on 26 September, 1940 the Woolston works received serious damage. Spitfire production there was brought to a standstill, and on the orders of the Ministry of Aircraft Production, Supermarine embarked on a dispersal programme. The company's headquarters were transferred to a stately country mansion in Hursley Park, near Winchester, and the design and experimental departments to accommodation in the spacious grounds surrounding the house. Final assembly continued at Eastleigh until November when the factory moved to Worthy Down, Hampshire, and gradually workshops were established in various other places in southern England within a 50-mile radius of Southampton.*

Meanwhile, in June 1940, the large shadow factory at Castle Bromwich had become a second source of Spitfire production. Its first Spitfires were Mk IIs and some fought alongside the Mk Is in the later stages of the Battle of Britain. (Similarly, Mk II Hurricanes entered squadron service late in the Battle, but the conflict was nevertheless fought mainly by the Mk I versions of the Spitfire and Hurricane.) Almost identical to the Mk I,

For further details see the author's monograph Supermarine Spitfire Remembered *(Vintage Aviation Publications, 1976).*

the Spitfire II had the 1,175hp Merlin XII engine using 100 octane fuel and fitted with a Coffman cartridge starter. A Rotol three-blade constant-speed propeller was standard and, compared to the Mk I with a similar propeller, the Mk II had a marginally improved climb rate and ceiling.

Spitfire IIs began to enter squadron service—with No 611 Squadron at Digby—in August 1940. The first 750 aircraft had A wing armament while the last 170 incorporated type B wings as a result of the new system of feed and shell case ejection for the Hispano cannon having been perfected. By 1 April, 1941, 650 Mk IIAs had been delivered and altogether some 57 squadrons were eventually equipped with the Mk IIA/B. In addition to the 920 true production aircraft a few more Mk IIs were converted from Mk Is, using surplus or reconditioned Merlin XIIs.

From December 1940 Spitfire IIs were frequently used in Fighter Command's early offensive sweeps in daylight over occupied Europe. These were either made in small numbers by the fighters themselves, when the operations were known as "Rhubarbs"; or in considerable strenghth accompanied by a few bombers in operations known as "Circuses". In each case, targets of opportunity were attacked in the hope of drawing enemy fighters into the air. The first Rhubarb was flown by Spitfire IIAs of No 66 Squadron, from Biggin Hill, on 20 December, 1940. The first Circus was flown by a Blenheim IV squadron escorted by six Spitfire and three Hurricane squadrons on 10 January 1941.

Several Spitfire IIs were fitted, in 1941, with a 40 gallon (182 litre) fixed external fuel tank on the port wing to allow them sufficient range for Circus operations against the Brest peninsular. These machines served with Nos 66, 118 and 152 Squadrons, but the bulky tanks made their handling characteristics extremely poor. One pilot of No 66 Squadron described them as "Bloody dangerous aeroplanes, very slow", and in fact several fatal accidents occurred as a result of pilots being unable to keep them on an even keel.

To meet a requirement in the air-sea rescue role, a Service modification of the Spitfire II appeared in 1942. This had a small rack for two smoke bombs under the port wing, inboard of the oil cooler, and two flare chutes in the fuselage, just behind the cockpit, housing a small dinghy and a metal food container. About 50 Spitfires were so converted and they served in several ASR squadrons (all of which were controlled by Fighter Command) and operated in conjunction with Lysander and Walrus ASR aircraft. The designation Spitfire IIC was adopted for this variant because it was, at the time, the logical means of distinguishing it from the existing Mk IIA and IIB models. Later, however, when role prefixes were introduced, the IIC was redesignated ASR II to avoid confusion with the Spitfire VC and subsequent models in which the C suffix indicated the so-called universal wing capable of mounting either A or B armament, or four 20mm cannon.

Not long after re-equipment of the squadrons of

Fig. 18 *An instructor at an Operational Training Unit demonstrates the Spitfire's retractable undercarriage and flaps to a group of fighter pilots undergoing their final training in June 1941.*

Fig. 19 *Sole external distinguishing feature of the Spitfire II was the small blister over the Coffman cartridge starter on the starboard side of the nose immediately behind the propeller, as seen on this IIA of No 65 Squadron.*

Fig. 20 *Armourers at a fighter station in Southern England remove and clean a Spit II's Browning guns after a sortie.*

Fighter Command with Spitfire IIs had been completed, re-equipment with Mk Vs began. The last front-line squadron to fly Mk IIs was probably the famous "Treble One" which, like certain other Spitfire units, was employed during the winter of 1941/42 on night fighting—a role for which the Spitfire was not suited due, amongst other things, to its narrow track undercarriage which always made things difficult for the pilot, even in daylight. Treble One Squadron relinquished the last of its Mk IIs in August 1942, at Kenley, having flown them alongside Mk Vs since the previous May.

Following their withdrawal from the squadrons, Mk IIs continued to serve in various units, chiefly at home, and, at first, mainly OTUs. Late in the war they could be found in such units as the Central Gunnery School and Technical Training Command Communication Flight. Others were used by the Air Fighting Development Unit to form instructional teams in air-fighting techniques—also known as Circuses but quite different from the operational ones—which were attached to various groups of Bomber and Coastal Commands and also to the Second Tactical Air Force.

SPECIFICATION

Spitfire IA
Powerplant: One 1,030hp Rolls-Royce Merlin II or III 12 cylinder liquid-cooled Vee engine.
Dimensions: Span 36ft 10in (11227mm); length (thrust line horizontal) 29ft 11in (9119mm); height over airscrew disc 12ft 3in (3734mm); wing area (gross) 242sq ft (22,5sq m).
Weight (normal loaded): 6,200lb (2812kg).
Performance: Max speed 362mph (583km/h); rate of climb 2,530ft/min (771m/min); combat range 395 miles (636km); ceiling 31,900ft (9723m).
Armament: Eight .303in (7,7mm) Browning machine-guns with 300 rounds per gun.

Spitfire IIA
Powerplant: One 1,175hp Rolls-Royce Merlin XII 12 cylinder liquid-cooled Vee engine.
Dimensions: Same as for Mk IA.
Weight (normal loaded): 6,275lb (2846kg).
Performance: Max speed 370mph (595,5km/h); rate of climb 2,600ft/min (7925,5m/min), combat range 395 miles (636km); ceiling 32,800ft (9997m).
Armament: Eight .303in (7,7mm) Browning machine-guns with 350 rounds per gun.

Fig. 21 *Ground crews load a dinghy and a food container into the chutes in the belly of Spitfire IIC P8131 AQ-C of No 276 Air-Sea Rescue Squadron early in 1943.*

Fig. 22 *Two of No 92 Squadron's trial cannon/machine-gun Spitfire IBs (R6908 QJ-F and X4272 QJ-D) take off from Manston, Kent, in company with an unmodified Mk IA (X4561 QJ-B) in December 1940.*

Fig. 23 *This "civilianised" Spitfire II (ex P8727) was re-engined in 1946 with a 1,440hp Merlin 45 for a private owner. Painted black and cream, and named "Josephine", it crashed on take-off from Kastrup, Copenhagen in April 1947.* [E. J. Riding.]

Fig. 24 *Cockpit of the Spitfire I.*

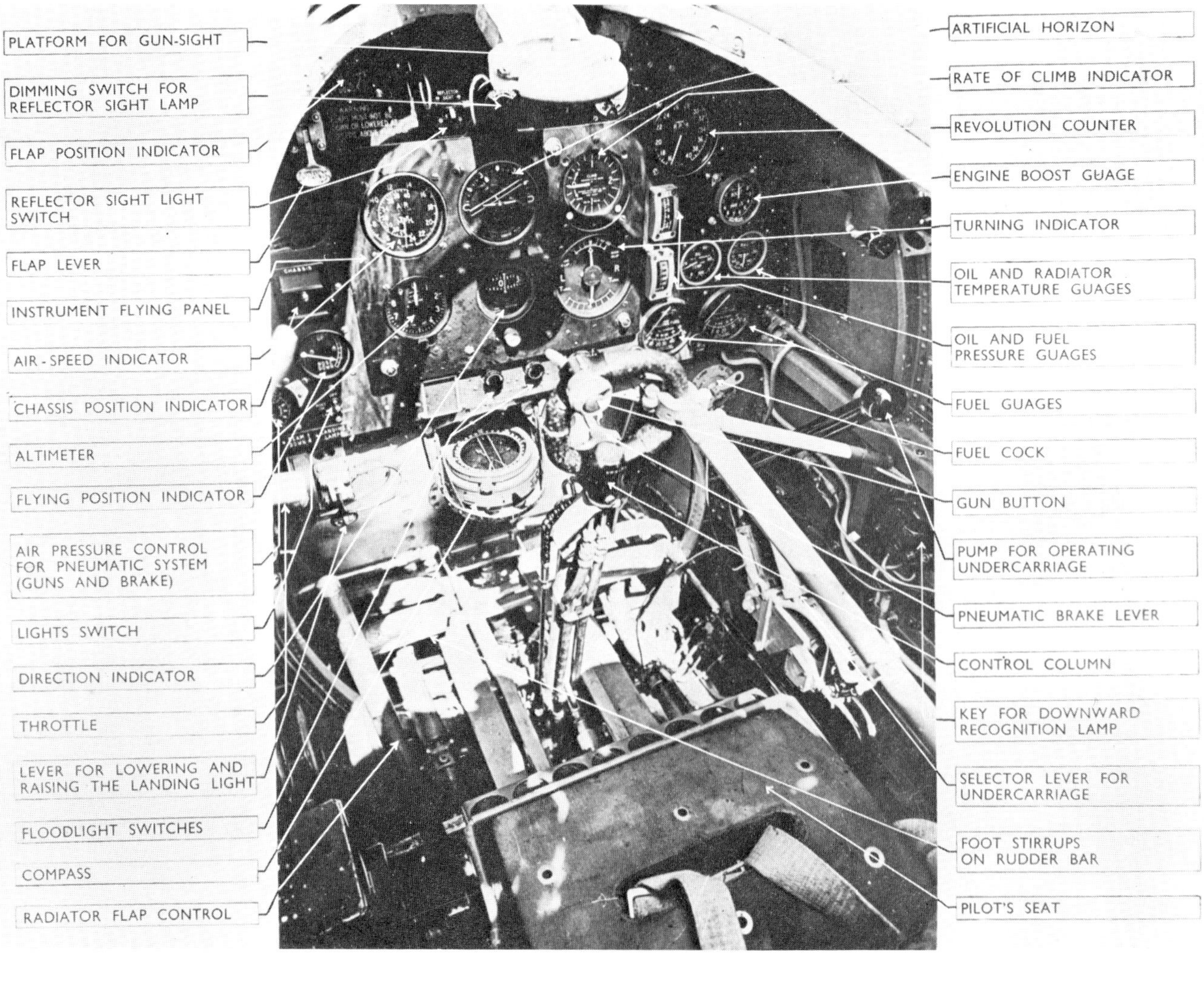

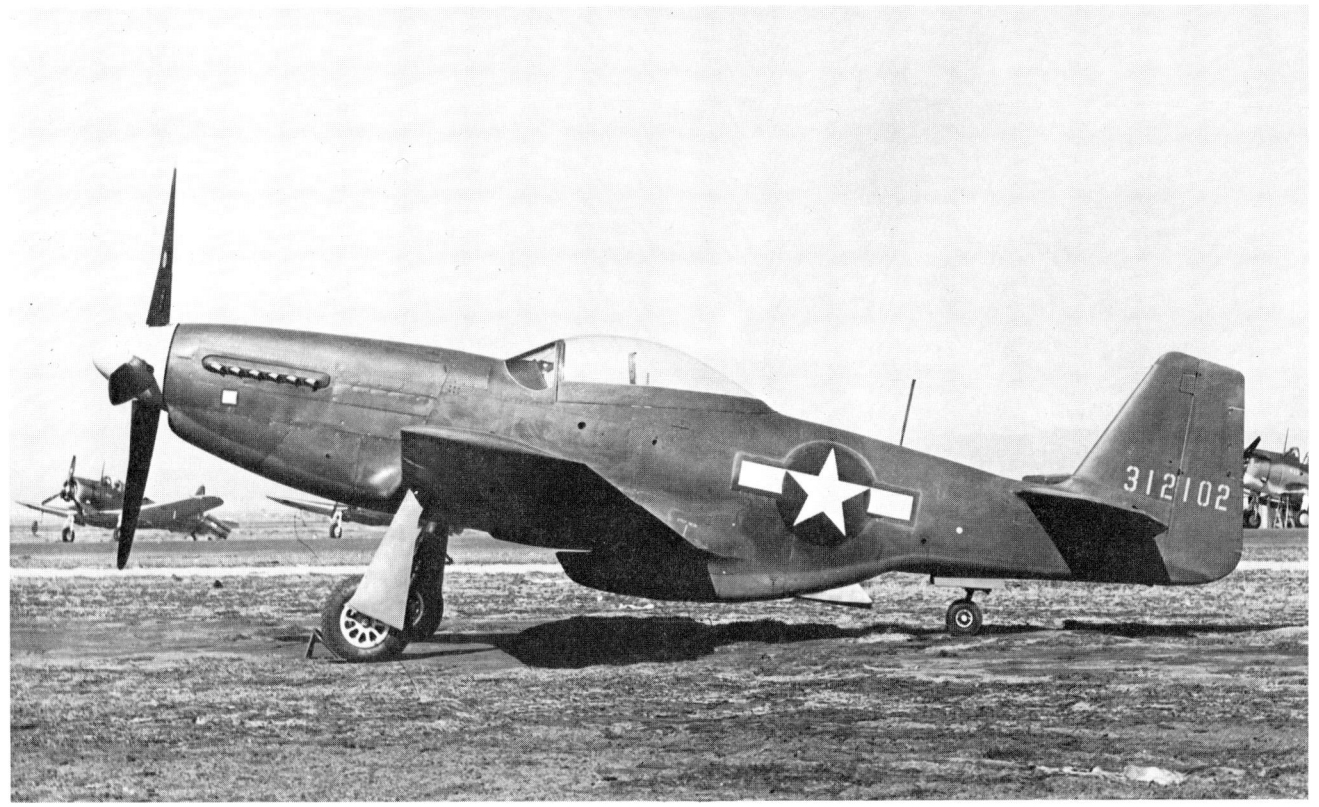

Fig. 1 *The tenth production P-51B-1-NA pictured in March 1943 after being re-engineered into the prototype P-51D.*

NORTH AMERICAN P-51D MUSTANG

By Harry Holmes

The last months of 1943 were crucial for the US 8th Air Force's strategic bombing of Germany as losses of both B-17 and B-24s were reaching an unacceptable level. These losses, mainly due to the effectiveness of the Luftwaffe fighter attacks, produced an urgent need for a long-range fighter escort to protect the bombers on missions far into Europe from their bases in England.

Escort duties were nothing new to the P-38 and P-47 equipped fighter groups of the 8th Air Force, but raids on targets deep inside Germany were now being made and the bombers were having to go it alone during the actual attack because of the limited range of the US fighters. In July 1943 long-range tanks had been fitted to the P-47 Thunderbolt, boosting the aircraft's radius of action to nearly 350 miles, but even this only afforded the bombers protection to the areas around the German border; and the P-38 Lightning, with a slightly longer range, was suffering technical troubles including frequent engine failures at high altitude.

In October 1943 the answer came with the arrival in England of the first production North American P-51B Mustangs. The P-51B model was the Merlin-powered version of the Mustang which had first flown three years previously powered by an Allison engine and, thus equipped, had served with both the USAAF and the RAF on tactical and army co-operation duties. The marrying of the basic Mustang airframe to the famous Rolls-Royce Merlin engine exceeded all expectations and turned the P-51 from just a good aircraft into what was arguably the greatest piston-engined fighter of World War 2.

Operations with the P-51B commenced in December 1943 when the 354th Fighter Group, a 9th Air Force unit, which, at that time was under the operational control of the VIII Fighter Command, carried out an uneventful fighter sweep over the French and Belgian coasts. This mission was led by Major Don Blakeslee of RAF Eagle Squadron fame "borrowed" for the occasion from his own unit, the 4th Fighter Group based at Debden.

The arrival of the P-51Bs in England proved timely as the 8th Air Force's bomber offensive got into full swing and the new fighter was proving to be an excellent aircraft. Most of the leading fighter aces built up their scores on this model, with famous names like Gentile, Godfrey, Beeson, Goodson and Hofer all racing into double figures during the spring of 1944. One of the well-known drawbacks of the P-51B was the poor visibility afforded the pilot due to the flush "razorback" cockpit arrangement of the aircraft, and rearward vision was exceptionally bad. The increasing number of combat missions flown by Mustangs made this problem more acute and North American's technical representatives in England studied the matter at some length before

Fig. 2 *Colonel Donald J. M. Blakeslee, commander of the famous 4th Fighter Group at Debden, Essex, was one of the finest fighter tacticians of World War 2.*

designing a clear-view bulged canopy similar to the sliding hood fitted to the Spitfire IX. Manufacture was commenced by the British engineering company of R. Malcolm Limited, and, known as the Malcolm hood, the canopy was fitted to most of the Mustangs serving in the UK with both the British and American forces including the P-51Cs which were entering service about that time.

Back in the United States the visibility problem had also been the subject of North American Aviation design team studies and a number of canopy types were looked at before it was decided that a "teardrop" shaped hood then being fitted to Britain's Hawker Typhoon would be the ideal solution. The tenth production P-51B-1-NA was taken from the assembly line at Inglewood and removed to the experimental department for the modification, which involved reducing in depth all the fuselage formers aft of the cockpit to enable the new canopy to be fitted. Wind tunnel tests with a one-third scale wooden model were carried out while the aircraft was being modified and flight tests on the completed aircraft brought to light only a few problems; as expected, the clear all-round view through the new canopy was excellent.

The success of these trials, which took place in February 1943, resulted in the allocation of a further two uncompleted P-51Bs to be used as manufacturing prototypes for the new model which was now known as the P-51D with the factory number NA-106. The P-51D would have two more .5in (12.7mm) wing guns than the P-51B/C models, increasing the armament to six machine-guns with an additional 620 rounds of ammunition for a total of 1,880. A strengthened wing allowed two bomb racks each capable of carrying a 1,000lb (454kg) bomb to be fitted and the power plant would be the new Packard-Merlin V-1650-7, known to the British as the Merlin 68.

The first order for the P-51D was placed on 13 April, 1943 authorising production of 2,500 aircraft. Later that month, on the 23rd, an order for 100 sets of Mustang components was received from the Australian government; these were subsequently assembled by Commonwealth Aircraft Corporation under a licence agreement with North American. The first four Australian-built Mustangs were P-51Bs while the re-

Fig. 3 *A new P-51D-10-NA on a production test flight from Inglewood, California, in August 1944.*

Fig. 4   *A coolant leak forced this 4th FG P-51D to crash-land in Germany on 12 September 1944—on only its second mission.*

maining 96 were P-51Ds and known by the Australians as CA-17 Mustang Mk XXs.

At the North American plants in Inglewood, California and Dallas, Texas, production of the P-51D was soon well under way and the last B and C models entered squadron service during the spring of 1944. The intensive fighting in the air over Europe made this theatre of operations the obvious choice for the introduction of the new model and May 1944 saw the first arrivals of the P-51D in England. It was with great excitement that pilots gathered around the P-51Ds when they reached the various bases throughout East Anglia. However, some of the excitement died when numerous check flights showed that the new aircraft did not perform quite as well as their current Mustangs, being slightly slower in the climb and in level flight. Directional control was poorer due to the lack of keel area on the cut-down rear fuselage and the addition of a dorsal fin fillet was found necessary. It was August 1944 before modification kits of the fin fillet became available and by that time 650 P-51D-5-NAs had already been delivered requiring the modification to be carried out in the field. Later aircraft would have the fillet added during manufacture and all P-51Ds built after 44-13902 were fitted with it on the production line.

Early operations with the P-51D showed the need for

Fig. 5   *This P-51K-5-NT, 44-11818, is seen just after roll-out from the North American plant in Dallas, Texas, in the summer of 1944.*

Fig. 6 *Major Henry Billie's P-51D "Prune Face" of the 355th FG, Steeple Morden, Herts, being prepared for a mission.*

Fig. 7 *"Rough and Ready" of the 355th FG displays a single victory symbol on the canopy frame.*

Fig. 8 *The 108 US gallon (409 litre) "paper" drop tank manufactured in England for the 8th Air Force.*

Fig. 9 *Lt Col Andrew J. Evans' mount "Little Sweetie 4" at Leiston, Suffolk, in April 1945. The fuselage star carried a red centre spot reminiscent of pre-war US markings.*

other modifications to be carried out, mainly electrical and radio changes. The installation of an MN-26C radio compass was undertaken at the same time as the fin fillet work. The strengthened wing had prompted the testing of various types of weaponry mainly due to the increasing amount of ground attack missions as well as "targets of opportunity" attacks on the way back from escort duties and an assortment of bombs, rockets, etc, were all tested. Perhaps the most significant of these weapons was the Bazooka type rocket tube, three of which were fitted under each wing, and zero-length rocket rails later became a standard fit on some later P-51s. A new-shape canopy also began to appear from the start of the P-51D-25 blocks and was easily identified by the noticeable bulged top behind the front frame of the sliding hood.

The large contracts placed with North American for the P-51D had the factories working all out and the production rate was such that Hamilton Standard, who at that time were manufacturing propellers for practically every US military aircraft, could not meet their quota for delivery of propellers for the P-51D. This crisis forced North American to look to other sources for the supply of propellers, amd so the P-51K was born. The P-51K was basically a D model which used an Aero-products propeller fitted with a new blade pitch changing mechanism and had hollow steel blades instead of the Hamilton Standard's solid aluminium type.

Fig. 10 *Soldiers of the US Army's 102nd Division discovered this P-51D, previously owned by the 352nd FG, when they captured the German town of Erding in March 1945.*

Fig. 11 *Although the war had ended, this P-51D, 44-14315 of the 364th FG, seen after a crash-landing at Honington, Suffolk, on 18 June 1945, resembles many wartime casualties which resulted from enemy action.*

The manufacture of the P-51K was allocated to the Dallas plant and after a faultless production record with earlier Mustangs things began to go wrong when the first of the K models were flight tested. Blade imbalance in the new propeller caused heavy vibrations in the aircraft and many of the production machines had to be rejected, the failure rate being as high as 19 per cent in September 1944. Eventually the vibration problems were brought to acceptable levels and by the time production ended 1,337 P-51Ks had been built including 597 which went to the RAF as Mustang IVAs.

With the arrival of the new P-51Ds into squadron service a number of the older B models were released for other duties including photographic reconnaissance with the designation F-6, although usually the title of P-51 persisted. It was decided in the middle of 1944 that PR versions could be built on the production line and the Dallas factory undertook the work, manufacturing F-6D and F-6K models which could be fitted with any of nine different types of camera in the K17, K22 or K24 series.

In 1944 ten two-seat conversions of the Mustang were built as TP-51Ds, this work also being undertaken by the Dallas plant. These aircraft had full dual control

Fig. 12 *One RAF operator of the Mustang IVA was 112 Squadron, whose famous shark's mouth insignia continued into the jet age. This particular aircraft, KH745, was once the personal mount of Gp Capt B. A. Eaton with the codes BA-E.*

Fig. 13 *Invasion-striped P-51D-5-NA "Tika-IV" of the 361st FG reveals its under-surface detail as it is caught by the camera during a vertical bank.*

and the second pilot was located under the original canopy. Fifteen aircraft were converted by Temco on the same principle after the war but used a slightly enlarged hood.

Of the many series of tests devised for the Mustang during its career one of the most interesting must be the deck landing trials on board the aircraft carrier USS Shangri-La (CV-38) during 1944. These trials had been requested as early as May 1943 and a P-51D was later allocated for the task with the aircraft leaving the Inglewood production line in February 1944. The installation of the arrester hook took longer than expected due to the extensive strengthening required in the rear fuselage, and by the time the trials were able to commence strategy in the Pacific campaign had changed, the US Navy's Hellcats and Corsairs coping admirably with the Japanese. The operations of the Mustang from the carrier proved to be a complete success and showed the aircraft's versatility once again. Despite reports to the contrary there was only one aircraft involved and this was confirmed by the pilot in charge, Lt Cdr (later Admiral) R. M. Elder, as a P-51D-5-NA, 44-14017, which was later handed over to the National Advisory Committee for Aeronautics (NACA).

As the war in Europe drew to a close the Mustang equipped every fighter group in the 8th Air Force with

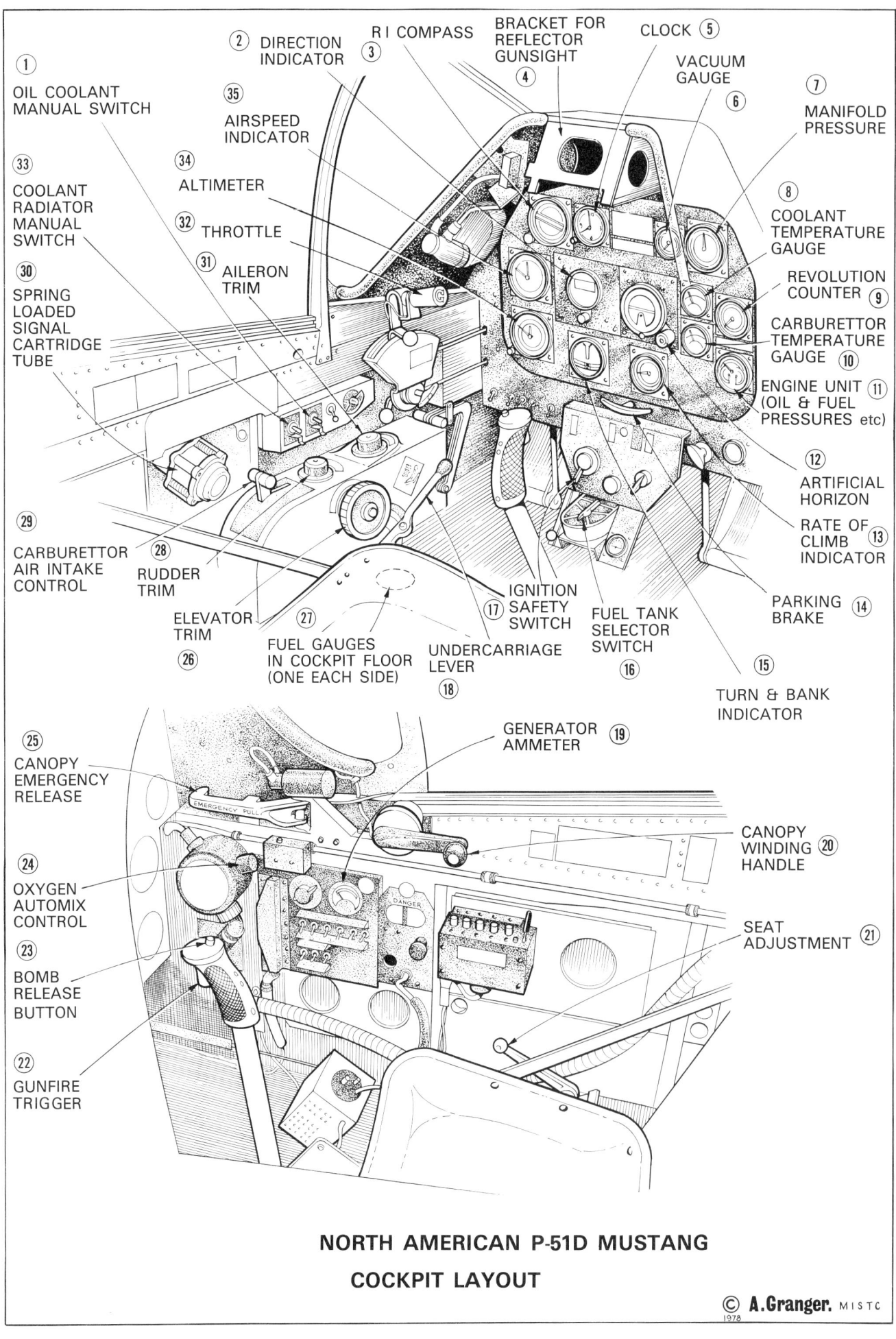

NORTH AMERICAN P-51D MUSTANG COCKPIT LAYOUT

P-51D OF
343 FIGHTER SQUADRON
55 FIGHTER GROUP
US 8 ARMY AIR FORCE

SCALE 1/72

RESEARCH: A. GRANGER © 1978
ARTWORK: ROY MILLS

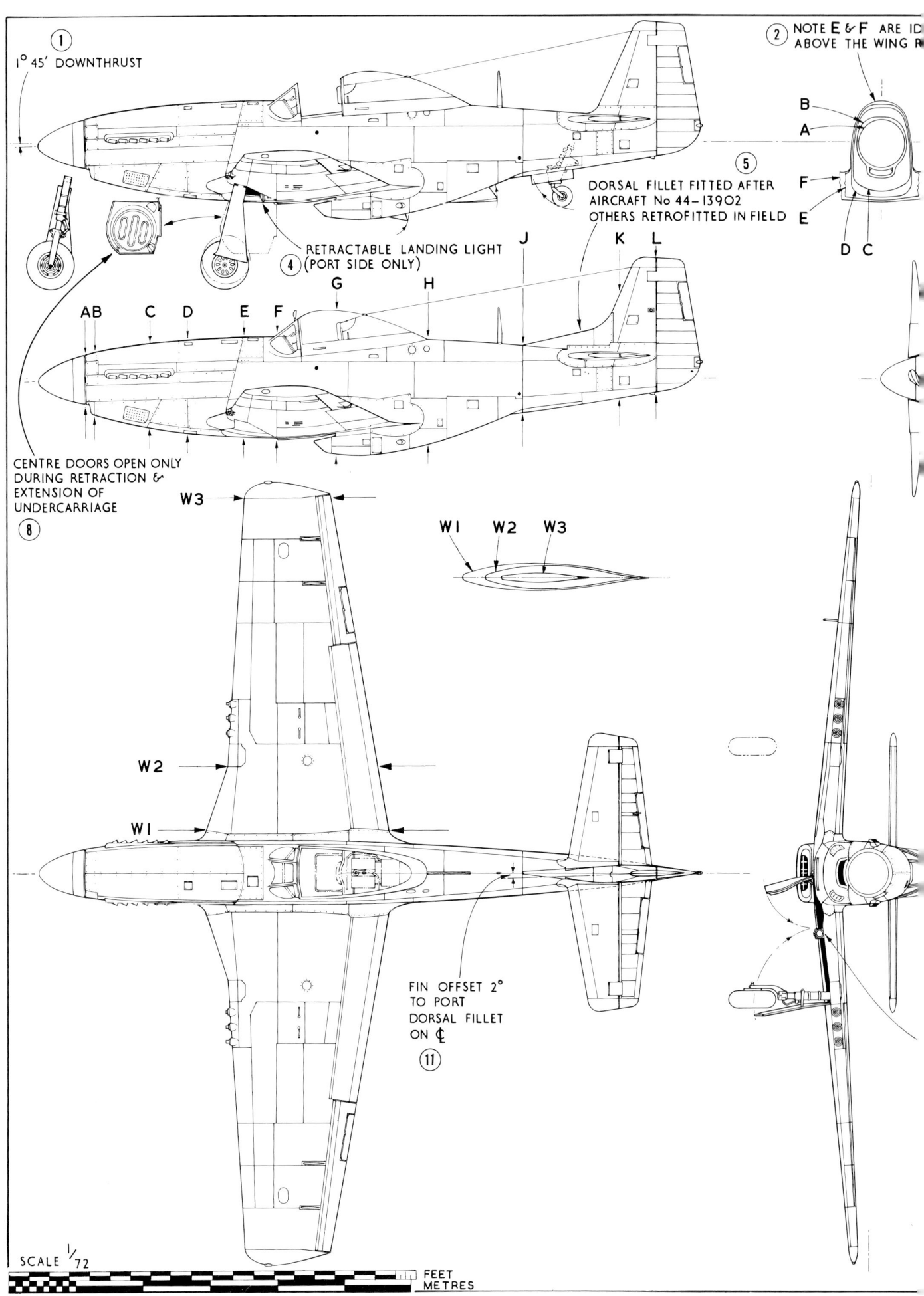

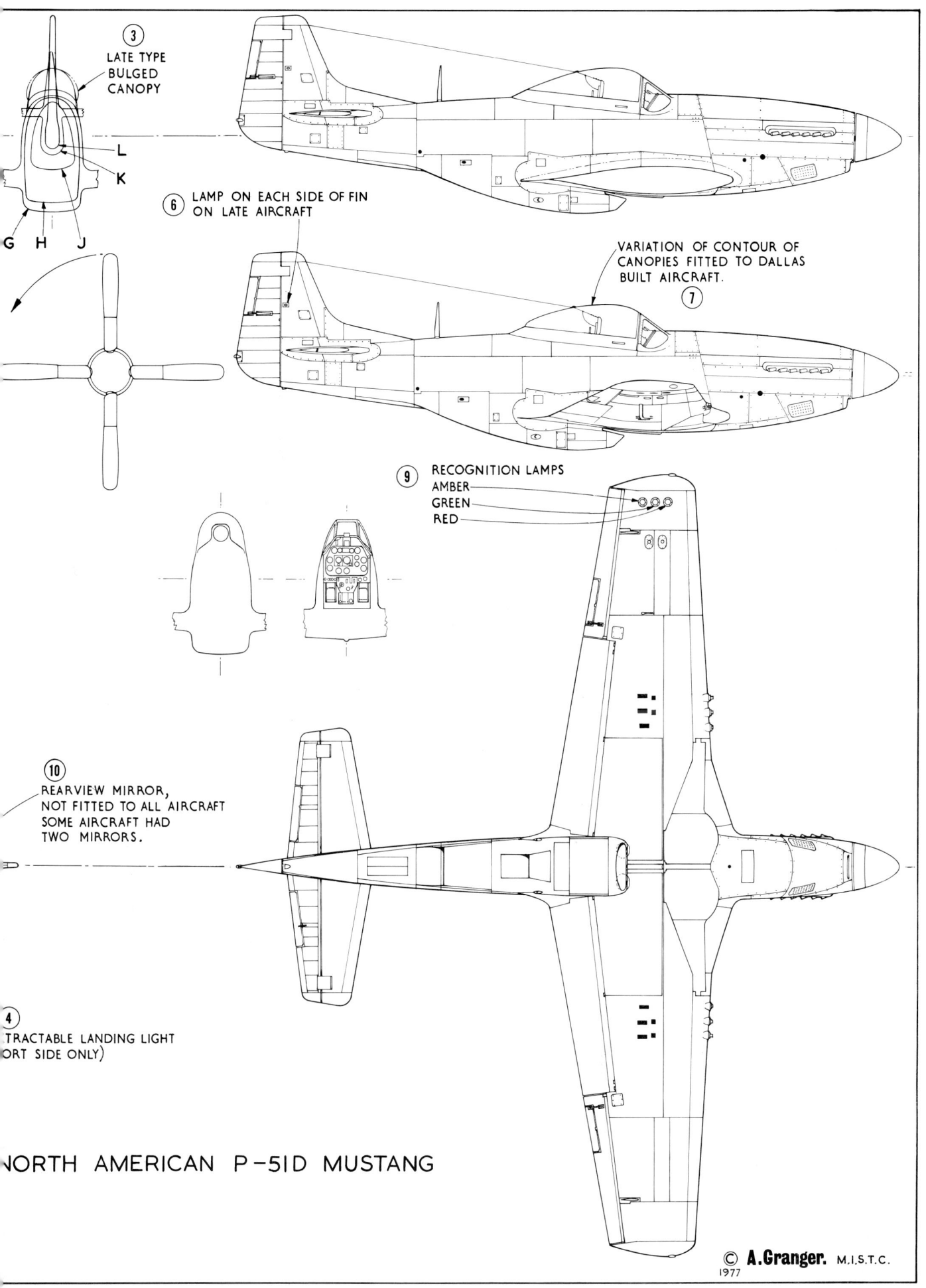

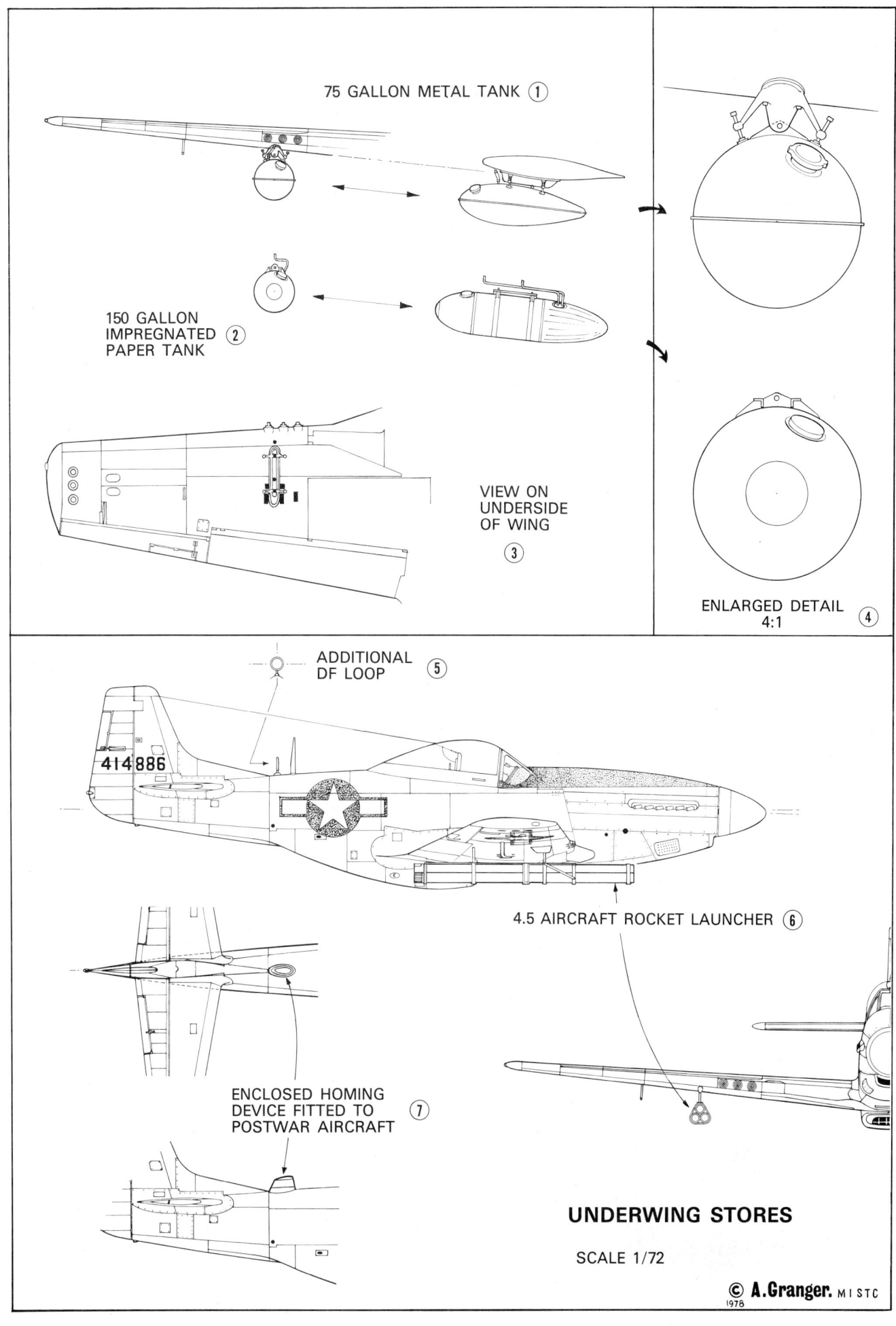

Fig. 14 *A flight of US-built P-51K Mustangs of 84 Squadron, Royal Australian Air Force, in 1945.*

Fig. 15 *One of the last batch of P-51Ds to be built at Inglewood, 44-74663 never saw action in World War 2, but looked ready to go when photographed in the Californian evening sunshine late in 1945.*

Fig. 16 *The Royal New Zealand Air Force received a batch of 30 P-51D-25-NTs at the end of the war, but these were placed in storage until the early 1950s when they were put back into service. NZ2423, alias 45-11513, was finally withdrawn from RNZAF service in June 1957.*

the exception of the 56th which steadfastly stuck to its P-47 Thunderbolts throughout its tour of duty in England. However, it is little known that the P-51 had indeed been selected for the 56th by its commander, Col Hubert Zemke, but this decision was reversed and in an interview with the author a number of years ago, Colonel Zemke told how it came about:

"In early 1943 it was known that the P-51 would be coming to the 8th Fighter Command. Knowing the superior range and air-to-air fighting potential of this aircraft to the P-47, I put in a bid for this aircraft for early re-equipment of the 56th Fighter Group. Though we were ahead of the 4th Fighter Group in the number of victories, the powers that be in Fighter Command stated that the 4th FG would be the first outfit in the 8th Air Force to get the Mustang while the 56th would be the fourth. I was satisfied with the situation at that time as there was plenty of fighting for all types of fighters in the late summer and fall of 1943. The fact remains that I felt, and still do, that the P-51 was better for aerial combat.

"Soon after, I was sent on a six-week bond tour of the United States and during my absence Col Robert Landry was assigned as Group Commander of the 56th. While I was away Col Landry, Col Robert Burns of 8th Fighter Command Headquarters and the 56th's Executive Officer Lt Col Dave Schilling reversed my decision on the P-51 and stated that the P-47 would be retained indefinitely by the 56th. Their contention for reversing my decision was that the P-47 was to receive a

Fig. 17 *After WW2, military aircraft in the United States were allocated "buzz numbers" to aid identification. These numbers consisted of a large two-letter code and the last three digits of the aircraft's serial. The Mustang was coded PF and the application can be seen in the photograph.*

'paddle blade' propeller and 400 more horsepower with the addition of water injection.

"Having flown the P-51 while in the United States and found it an excellent airplane, I was livid with rage upon my return to England to find that I had been sold down the river. It did not help when I learned that the decision had been approved by General Kepner at 8th Fighter Command Headquarters.

"As the spring of 1944 went by the Germans were concentrating their fighter forces deeper in Germany and our P-47s, through lack of range, could only escort the bombers into France or only just inside Germany and we also took the 'milk runs' whereas the P-51 groups could fly deep into Germany where the 'hunting' was much better.

"When the Commander of the 479th Fighter Group was posted missing I was asked to give up Dave Schilling to take over as the new commander, but Dave refused to go as the 479th had a poor record at that time. As that outfit was due to change its P-38s for P-51s in August 1944 I called General Kepner and said that I would be delighted to take over the 479th as it would be a challenge and it would also give me a chance to fly the P-51 again and perhaps prove that it was a superior airplane."

Colonel Zemke did take over the 479th and with the arrival of such a distinguished airman as the new commander and re-equipping with the Mustang soon after, the group's morale began to soar. By the time the war ended the 479th had a respectable total of victories. Unfortunately, Hub Zemke could not take all of the credit for this turnaround in fortunes because in October 1944 his P-51D broke up in severe turbulence during a combat mission over Germany and Zemke spent the duration in a prisoner of war camp. However, perhaps the introduction of the P-51 into the 479th had something to do with it.

During the last year of the war Mustangs roamed the skies over Europe with the now standard 108 US gallon (409 litre) compressed paper drop tank giving the

Fig. 18 *The Swiss Air Force's P-51D, J-2012, seen here, once served with the 370th FG, 9th Air Force, during World War 2.*

Fig. 19 *The nearest of these California Air National Guard P-51Ds of the 195th FG is seen to be carrying small practice bombs.*

aircraft an 850 mile radius of action. Various new innovations were tried during this period with introduction of the APS-13 tail warning radar being probably the most interesting, although this did not see service in great numbers. German jet and rocket fighters were met on numerous occasions by P-51s of both the USAAF and RAF and proved that, if conditions were right, the piston-engined fighter could hold its own against the Me 262s and Me 163s of the Luftwaffe.

As World War 2 ended P-51s were not only active in Europe, but in the Mediterranean and the Far East, where in the war against Japan the aircraft's extensive range again allowed bombers on long missions to be escorted all the way to the target; this time they were B-29 Superfortresses bombing the Japanese mainland from their island bases in the Pacific.

In the final analysis of enemy aircraft destroyed it was proved that the P-51 could claim 13 victories for every hundred sorties flown while the P-38 and P-47 trailed way behind with four and three victories per 100 sorties respectively.

With the war's end, contracts for warplanes were reduced drastically or cancelled completely, and orders for the Mustang were no exception: 1,000 P-51Ds were cancelled and of the 2,000 new P-51H model on the Inglewood production line only 555 were completed. Orders for 1,700 P-51Ls which was a re-designed H model, and 1,628 P-51Ms were cancelled before production really got under way, although one P-51M, 45-11743, a re-engined P-51D (the last P-51D on the Dallas line) was completed.

The Mustang production lines, which were turning out more than 500 aircraft each month by January 1945, were quickly run down and the last of the type, a P-51H, was rolled out at Inglewood on 9 November, 1945. The plant at Inglewood had built 6,502 P-51D2 while Dallas had produced 1,454 P-51Ds, 1,337 P-51Ks plus 136 F-6Ds and 163 F-6Ks, the complete wartime production of the Mustang totalling 15,582 aircraft.

The arrival of peace saw some of the USAAF's front-line fighter squadrons relinquishing their P-51Ds for the new H model although the jet-propelled Lockheed P-80 Shooting Star was also beginning to enter service.

In May 1946 a US government act was approved to establish a reserve air unit to become part of the National Guard. The personnel would be civilians doing their normal job during working hours, but devoting their evenings and weekends to manning America's second-line air force. From the Air National Guard's establishment until December 1948, 28 fighter squadrons of the ANG received more than 700 surplus P-51Ds. These aircraft remained the backbone of the guard until they were gradually replaced or supplemented by P-51Hs in the early 1950s and by 1952 68 of the 98 ANG squadrons were operating Mustangs.

In July 1947 the Army and Air Force in the United States had become separate Services, changing the USAAF into the USAF. The following year the Air Force introduced a new designation system for the aircraft with P for Pursuit changing to F for Fighter, so the Mustang become the F-51. This also affected the photographic reconnaissance versions, the F-6D and F-6K becoming the RF-51D and RF-51K.

In June 1950 the North Korean Army crossed the 38th Parallel, thus invading South Korea, and in support of United Nations resolutions the United States, Britain, Australia and other member countries were pledged to assist the south and so the Mustang went to war once more. Air National Guard units were alerted and nearly 150 Mustangs were drawn from units all over the United States and flown to Alameda Naval Air Station in California for shipment to the war zone. At Alameda they were given a protective coating against the corrosive sea water and then loaded on to the carrier USS Boxer for transportation to Japan before

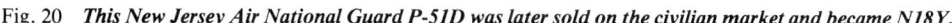

Fig. 20 *This New Jersey Air National Guard P-51D was later sold on the civilian market and became N18Y.*

Fig. 21 *To war once more. This F-51D of the South Korean Air Force (ROKAF), photographed at Kimpo in 1951, was ex 44-63581, an old 8th Air Force machine.*

Fig. 22 *Three Mustang Mk 4s from the Royal Canadian Air Force's Central Flying School, Trenton, in April 1952. The two aircraft nearest the camera have been equipped with tow-target racks.*

Fig. 23 *An Australian-built Mustang XXI, A68-104, photographed at Mascot, New South Wales, on 10 November 1961.*

Fig. 24 *The Reno Air Races always see a large turn-out of Mustangs, and here we see a P-51D-20-NA being prepared for one of the heats during September 1969. This machine, ex 44-63872 and RCAF 9552, was destroyed in an accident the following year.*

detachment to Korea itself. All of the aircraft selected for duty were F-51Ds as they were to be used mainly for ground-attack operations for which they were better suited than the lighter F-51H. The Mustangs were almost all late models with underwing attachment points for bombs or rockets.

As in World War 2 the Mustang acquitted itself admirably in Korea, flying in front-line service throughout the conflict, and in addition to the USAF, P-51s were operated by Australia and South Africa as well as the South Koreans themselves. When the Armistice was signed in July 1953 only one USAF Mustang unit was operational, the other F-51 squadrons being replaced by jets at various intervals.

Back in the USA Mustangs continued to serve in the Air National Guard until March 1957 when the last F-51D in the ANG was handed over to the USAF Museum at Wright-Patterson AFB, Ohio, for permanent exhibition.

However, that was by no means the end of the Mustang story as many other countries continued to operate the aircraft including Italy, Nationalist China, Sweden, New Zealand, Israel, Canada, France, Switzerland, the Philippines, Cuba, Somalia, Uruguay, Holland, Guatemala and Nicaragua. At the time of writing the machine still serves in small numbers with the air forces of Bolivia, Dominica, Haiti, the Honduras, Indonesia and Salvador.

Many surplus Mustangs were bought by Trans-Florida Aviation, which later became known as the Cavalier Aircraft Corporation, specialising in the conversion of the aircraft to two-seaters for the civilian market and also rebuilding and improving airframes for the military. Later developments by the company included the Mustang II which had an uprated Merlin, strengthened wings and tip tanks, combining range with extra load carrying capacity. The Cavalier Mustang III had its Merlin replaced by a Rolls-Royce Dart turboprop engine to improve performance and economy.

In the United States Mustangs can still be seen in large numbers, with the Reno Air Races and numerous fly-ins being the favourite venues for owners. The US Army has also been using Mustangs in recent years as chase planes for the trials of a number of high-speed helicopters such as the Lockheed Cheyenne and the Sikorsky Blackhawk.

The enthusiasm for the Mustang is still high and any appearing on the civil second-hand market are quickly purchased, even at inflated prices. The aircraft, which seems to be referred to as the P-51 once more, is actually increasing in numbers on the US civil register due to once grounded machines being made airworthy after many hours of rebuilding and ex-foreign air force P-51s being imported.

Today, 36 years after its birth, the Mustang is a living legend and it is hoped that many examples will be flying well into the next century, still proving to be the finest piston-engined fighter ever built.

SPECIFICATION

Powerplant: One 1,490hp Packard Merlin V-1650-7 driving a Hamilton Standard 4-blade 11ft 2in (3404mm) diameter constant speed propeller (D model) or an Aeroproducts 4-blade 11ft (3353mm) diameter constant speed propeller (K model). Internal fuel capacity was 85 US gal (322 litres) in fuselage tank and two 92 US gal (348 litres) tanks in the wings totalling 269 US gal (1018 litres).

Dimensions: Wing span 37ft 0 5/16in (11285mm); wing area 233.19sq ft (21.664sq m); length 32ft 3 1/4in (10446mm); height, overall 13ft 8in (4166mm).

Weights: Basic empty 7,125lb (3232 kg); max take-off 11,600lb (5262kg).

Performance: Maximum speed 437mph (703km/h) at 25,000ft (7620m); landing speed 100mph (161km/h); time to altitude 13.1 minutes to 30,000ft (9144m); service ceiling 41,900ft (12.771m); range (clean) 950 miles (1529km) cruising 362mph (582km/h) at 25,000ft (7620m); range (2 × 110 US gal (416 litres) external tanks) 2,300 miles (3701km) cruising 245mph (394km/h) at 10,000ft (3048m); range (2 × 108 US gal (409 litres) external tanks) 2,258 miles (3633km) cruising 245mph (394km/h) at 10,000ft (3048m).

Armament: Six .5 inch Browning machine-guns with amunition totalling 1,880 rounds. 400 rounds in each of the inboard guns with the centre and outboard each having 270 rounds. Two 1,000lb (454kg) bombs could be carried when the aircraft had no external tankage. With bomb racks fitted the aircraft could also carry six 5-inch High Velocity Aircraft Rockets (HVAR) or ten HVAR when without racks.

Acknowledgements: The author wishes to express his appreciation to the following sources for the use of photographs contained in this publication: US Air Force, US Navy, North American Aviation Inc, Public Archives of Canada, Royal Swedish Air Force, RAAF, Danny Morris, Bob O'Dell, Brian Goulding, John Hopton, Michael O'Leary and Lt Gen Andrew J. Evans USAF.

Fig. 25 *P-51D close-up showing the Hamilton Standard 4-blade hydromatic propeller of 11ft 2in (3403.6mm) diameter.*

Fig. 26 *P-51D undercarriage, air scoop and external stores rack.*

Fig. 1 *A "Schwarme" of Me 109E-3s of 7 Staffel, Jagdgeschwader 51.* [via Bruce Robertson.]

MESSERSCHMITT 109E
By Peter G. Cooksley

Fig. 2 *Me 109E-4 Trop Black 3 of I/JG27 in an unusual striped camouflage scheme at its base in North Africa.* [via Bruce Robertson.]

Fig. 3 *An early production Me 109E-3 with the radio call sign codes CE+BM on test before delivery to the Luftwaffe.* [via Bruce Robertson.]

Fig. 4 *First development aircraft for the E-series was the Me 109 V14, which began flight trials in the summer of 1938 and was eventually registered D-IRTT.*

It had been a frustrating day for the German pilots of Airfleet 2 had the long-awaited "Eagle Day", which heralded the concentrated series of attacks on targets in Great Britain. The day—13 August, 1940—had dawned with some early morning mist and slight drizzle, but this had cleared and although there was some cloud over the English Channel the day was set for mainly fair weather. The bombers had been away fairly early and "H-hour" had been put forward too late to recall them, so they proceeded without escort.

A great feeling of relief was therefore the overall emotion on all the fighter bases when, half way through the afternoon, the final order to begin operations was received. From Dinan, thirty Messerschmitt 109s of II/JG53 were swiftly heading towards the coast headed by Major Günter Freiherr von Maltzahan. One of the following fighters was piloted by Leutnant Heinz Pfannschmidt in an E-1 model. Scrambled to meet the raid and already airborne were the Spitfires of No 609 Squadron from Warmwell, one of which was flown by Pilot Officer D. M. Crook. Over Poole Harbour the two met; momentarily the Messerschmitt flashed into the Spitfire's reflector sight, Crook's fingers slackened the safety catch on the control column and pressed the firing button and the other fighter quite suddenly heeled over and plummeted to earth at Lyme Bay taking Pfannschmidt to his death. The time was just about ten past three and the first Messerschmitt to be sacrificed on Eagle Day had fallen.

Details of the only German single seat fighter to participate in the Battle of Britain were, however, already known to Great Britain. Two "Emils" had fallen into French hands late in 1939 and, being intact, had been the subject of tests at the Orléans-Bricy Test Centre. Here, one of them had crashed but Me 109E-3 Werke Nummer 1304, formerly of II/JG54, had been passed, in May, 1940, to the RAE at Farnborough where it was examined by three British pilots. It was discovered that the design suffered from aileron snatching as the automatic slots opened, which made the response heavy at slow speeds while some sustained physical effort was called for in their use at high speed; but the rate of climb was good and there was a gentle stall and a reluctance of the engine to cut out, both qualities being retained even under "G". Many pilots found the cockpit claustrophobic, partly due to the sideways-opening hood but mainly due to its cramped dimensions—a thing which inconvenienced pilots of both sides if they were tall and seated on a parachute pack.

The earlier Messerschmitt Me 109D had proved itself in the Spanish Civil War and the lessons learnt there had been incorporated in the Me 109E, some early models

Fig. 5 *Me 109E-3 of Major Schellmann of JG2 "Richthofen" partly stripped down after receiving battle damage in 1940.* [via Bruce Robertson.]

Fig. 6 *An early production E-3, Werke Nummer 1304, White 1 of II/JG54 "Grunherz", seen after falling intact into French hands on 22 November 1939 when its pilot mistakenly landed near Woerth, Bas-Rhin, some 12 miles on the French side of the border.*

Fig. 7 *After testing by the French at Orléans-Bricy, W/Nr 1304 came to Britain in May 1940 and, wearing British insignia, was evaluated by the A&AEE at Boscombe Down and the RAE at Farnborough.*   Fig. 8 *In June 1940 W/Nr 1304 was allotted the RAF serial AE479. It also flew with the Air Fighting Development Unit before being shipped, in April 1942, to the USA (with the tail unit of another Emil fitted following a crash-landing) for evaluation at Wright Field.*

Fig. 9 *Plan view of AE479, alias W/Nr 1304. This Emil was the first enemy aircraft to be evaluated in this country in WW2 and provided the RAF with valuable information about its chief adversary.*

of which also flew in that conflict, this in turn having been evolved from the Me 109V14, the true prototype of the Emil, which had first flown during the summer of 1938. The original model of the new fighter was designated the Me 109E-0 and this, the pre-production variant, was armed with only four MG 17s: a pair above the cowling synchronised to fire through the airscrew arc and one each in the outer wing panels.

It was on this model that some weakness was found in the undercarriage and there were some taxying accidents; but on the credit side it must be recorded that all pilots were agreed on the unusual steadyness while moving over rough or indifferent ground, despite the long legs and narrow track, a stability which tended to be weighted against the view from the cockpit while on the ground which fresh pilots described as "appalling".

All the early models were powered by the Daimler-Benz DB.601A, an engine fitted with a fuel injection pump instead of a carburettor so that there was no need to throw the machine on its back at the start of a dive as was the custom among contemporary British fighters, to obviate petrol starvation from the resultant "G" effect.

With the tendency to develop flutter under certain conditions and of which there had been dark rumours seized on by propagandists who coined the term "flutterschmitt" duly laid, production went ahead on the E-1 version which began to reach Luftwaffe fighter units during the spring of 1939. The first production models had the same armament as the evaluation E-0 type but the later machines were fitted with a 20mm MG FF in each wing instead of the twin MG 17s which were retained only above the fuselage.

Machines of this type also went to Spain, although too late to see operational service; but it was essential that the victories of General Franco's forces be followed up as swiftly as possible, so the bargain was struck that re-equipment of the Condor legion which had been allowed to run down should be carried out at entirely the Nationalist's cost, payment being made in kind by the supply of iron ore.

Fig. 10 *Looking rather sorry for itself is this brand new E-4 which suffered a mishap during testing.* [via Bruce Robertson.]

Spanish machines thereafter appeared with lead grey upper surfaces with light blue underparts which had distinguished the earlier 109-Ds which had similarly been marked with black discs with white crosses above and below the wings, while the colours were reversed on the rudder which was entirely white with a black cross saltire. Fuselage markings consisted of a wide black flash running aft from the forward exhaust, while a black disc served as a national insignia on the rear fuselage; this disc was prefixed by a type numeral and an individual number followed. For identification purposes some wing tips were doped white above and below, while spinners could be either the same shade as the fuselage or be finished black.

Spain was not the only country to fly the 109-E outside its country of origin since a number went also to equip Switzerland's Air Force. These were flown in a mid-green camouflage above and light blue below with the national marking of a white cross on a wide red band round the wing tips and the upper parts of the fuselage; inboard of this on the wings and aft on the fuselage appeared a series of alternate red and white bands with a white serial number, eg J-319, on the rear

Fig. 11 *Luftwaffe ground crew re-arm a Me 109E-3.* [via Bruce Robertson.]

fuselage. The entire rudder was red on these E1s with a small white cross on the upper portion, while noses of these fighters were finished entirely white as an additional indication of neutrality in the early months of the European war. After some ten years' service some of these Messerschmitts were not struck off charge until 1949.

Due to pressure on space brought about by the production in the Augsburg factory of the Me 110, work was moved to Regensburg and at the same time considerable sub-contracting was begun. It was from the new factory that there emerged the fighter-bomber version, the Me 109E-1/B fitted with differing racks for the delivery of either four 50kg bombs or a single 250kg bomb, there being a total of 850 E-1s on the Luftwaffe's first-line strength at the beginning of the Nazi attack on Poland. The following September when the direction of hostilities had changed it was the concept of this, the fighter-bomber version, which was to bring about the unpopular decision that one Staffel in each Jagdgeschwader should fly the model with bomb racks.

Before the end of 1939 there began to enter service the improved model designated the 109E-3, this being the type which saw some service in the Battle of France and was encountered in large numbers over the British Isles during the attacks in the following summer, a few months before which production was achieved at a rate of 150 per month. This type not only incorporated a change of engine to the DB.601Aa but also saw the introduction, on the earlier examples, of an MG FF gun to fire through the hub of the VDM electrically operated controllable-pitch airscrew. In later E-3s this 20mm weapon was deleted and the aperture utilised for a cooling inlet for the generator. It is interesting to note that one machine of this type was subscribed for and presented by Saar miners to Captain Molders, group Commander of JG51, the machine being marked with the insignia of these workers—two crossed hammers—on the cowling. There is some doubt if he in fact ever flew the aircraft in combat.

Meanwhile, an improved variant was entering the production stage which had a reduced weight and a major change in armament; this was the E-4 on which the hub gun, dropped from the earlier type after a time,

Fig. 12   *AE479, alias W/Nr 1304, shows off its sleek profile.*

Fig. 13   *A Me 109E-4 with manufacturer's radio call sign codes.*

67

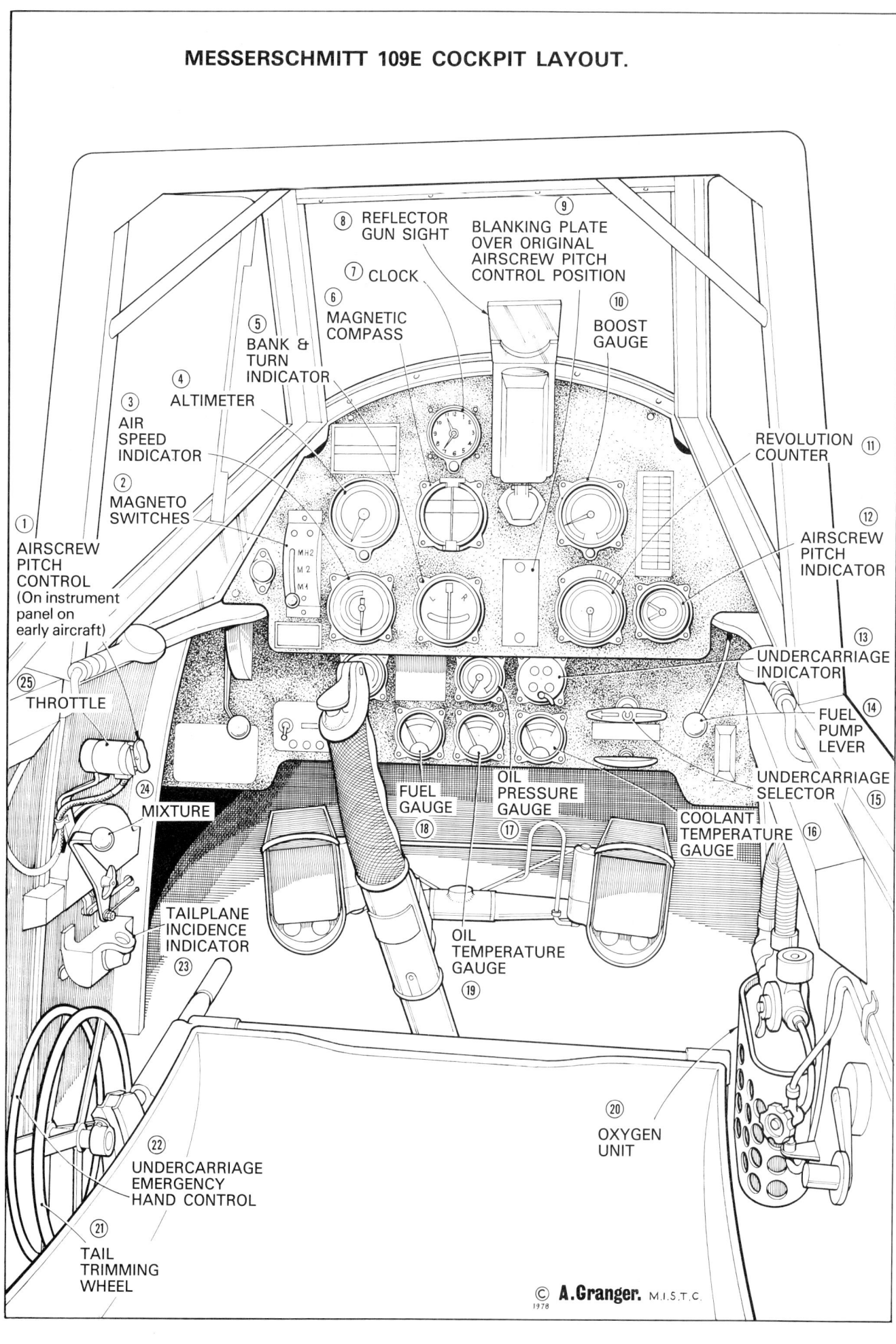

1. STANDARD 71 (DARK GREEN)/02 (GREY) CAMOUFLAGE ON TOP SURFACES WITH 65 (LIGHT BLUE) UNDERNEATH & ON FUSELAGE SIDES.

2. MESSERSCHMITT 109E-4. BROUGHT DOWN NEAR MARDEN, KENT ON SEPTEMBER 5, 1940, WHILE BEING FLOWN BY Oblt. FRANZ von WERRA, ADJUTANT, 11/JG3 "UDET". von WERRA LATER ESCAPED IN CANADA & EVENTUALLY RETURNED TO GERMANY.

3. NOTE REPLACEMENT PANEL IN FRONT OF WINDSCREEN, FINISHED IN 71 (DARK GREEN) MOTTLE ON 65 (LIGHT BLUE)

4. AIRSCREW 70 (BLACK GREEN) SPINNER 70 (BLACK GREEN) AND WHITE SEGMENTS WITH 02 (GREY) BACKPLATE.

SCALE 1/72

RESEARCH: A. GRANGER. © 1978
ARTWORK: ROY MILLS.

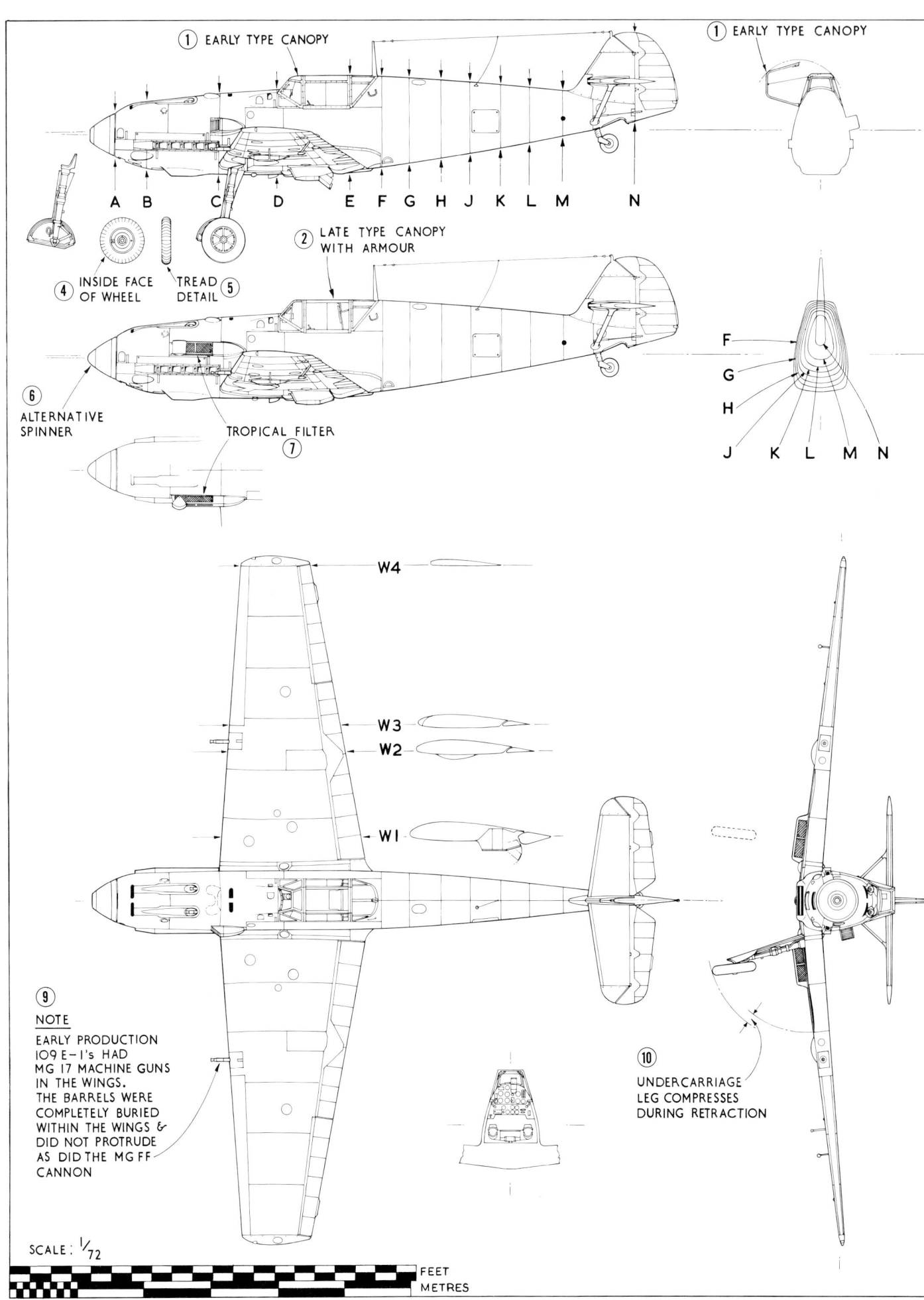

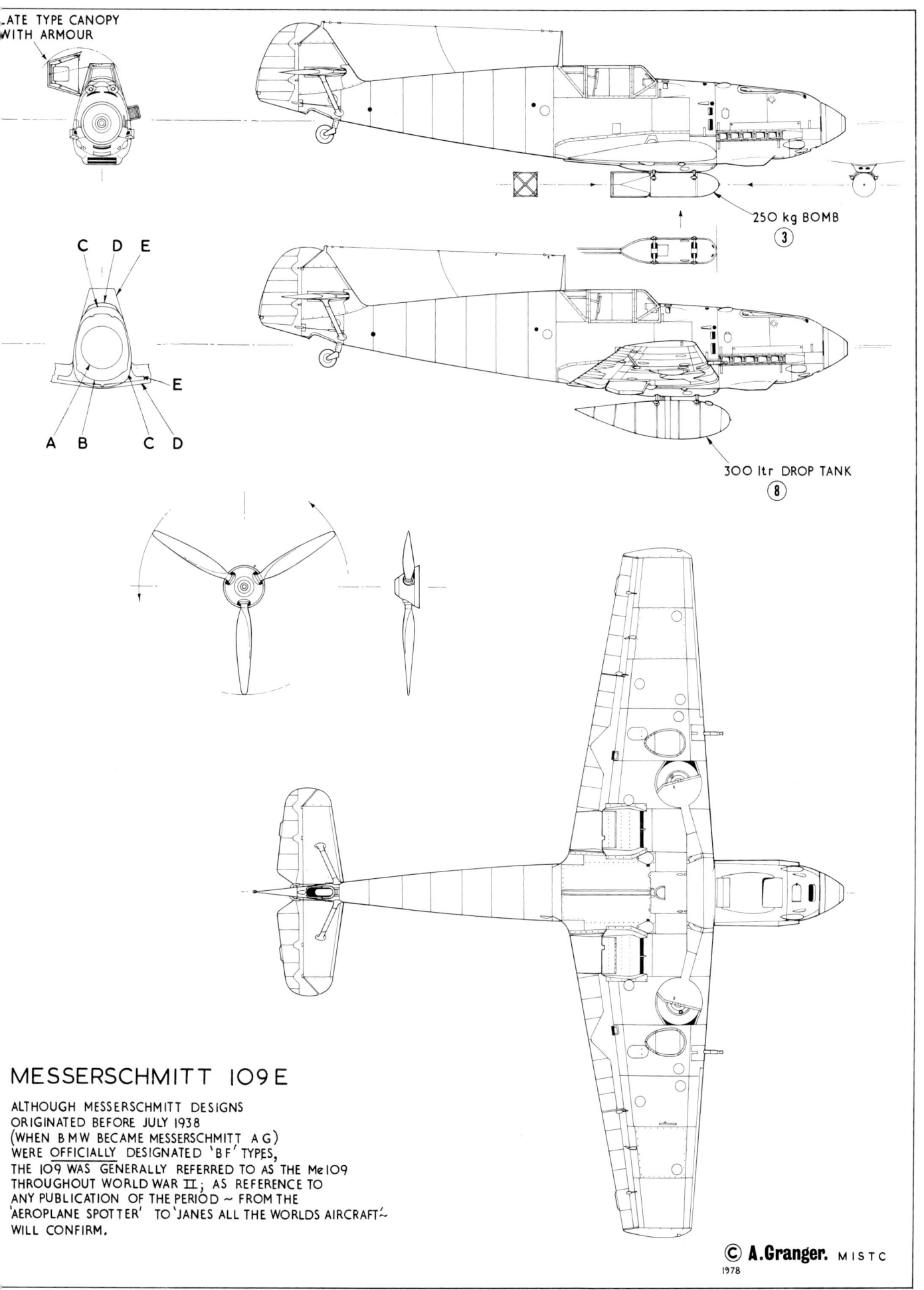

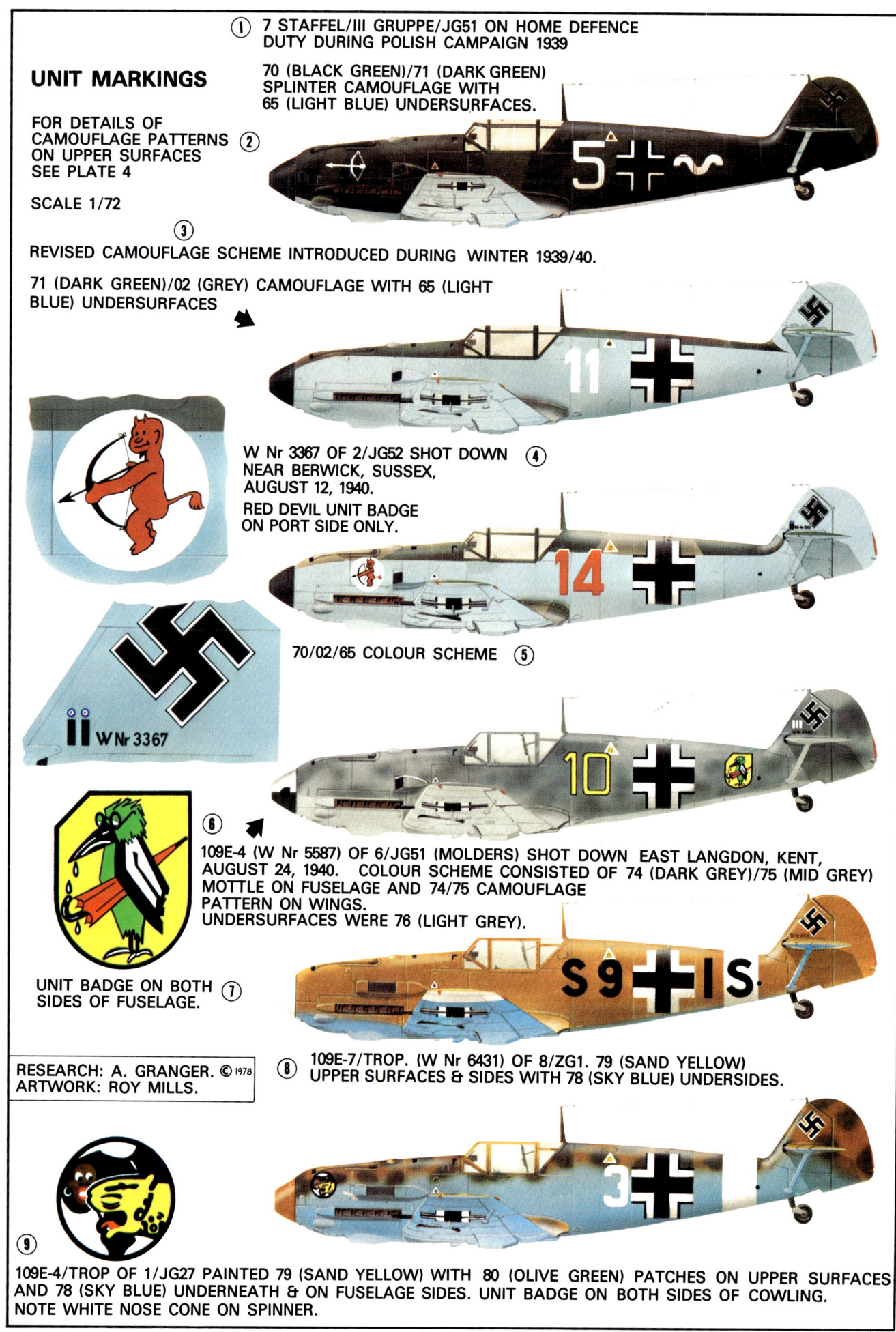

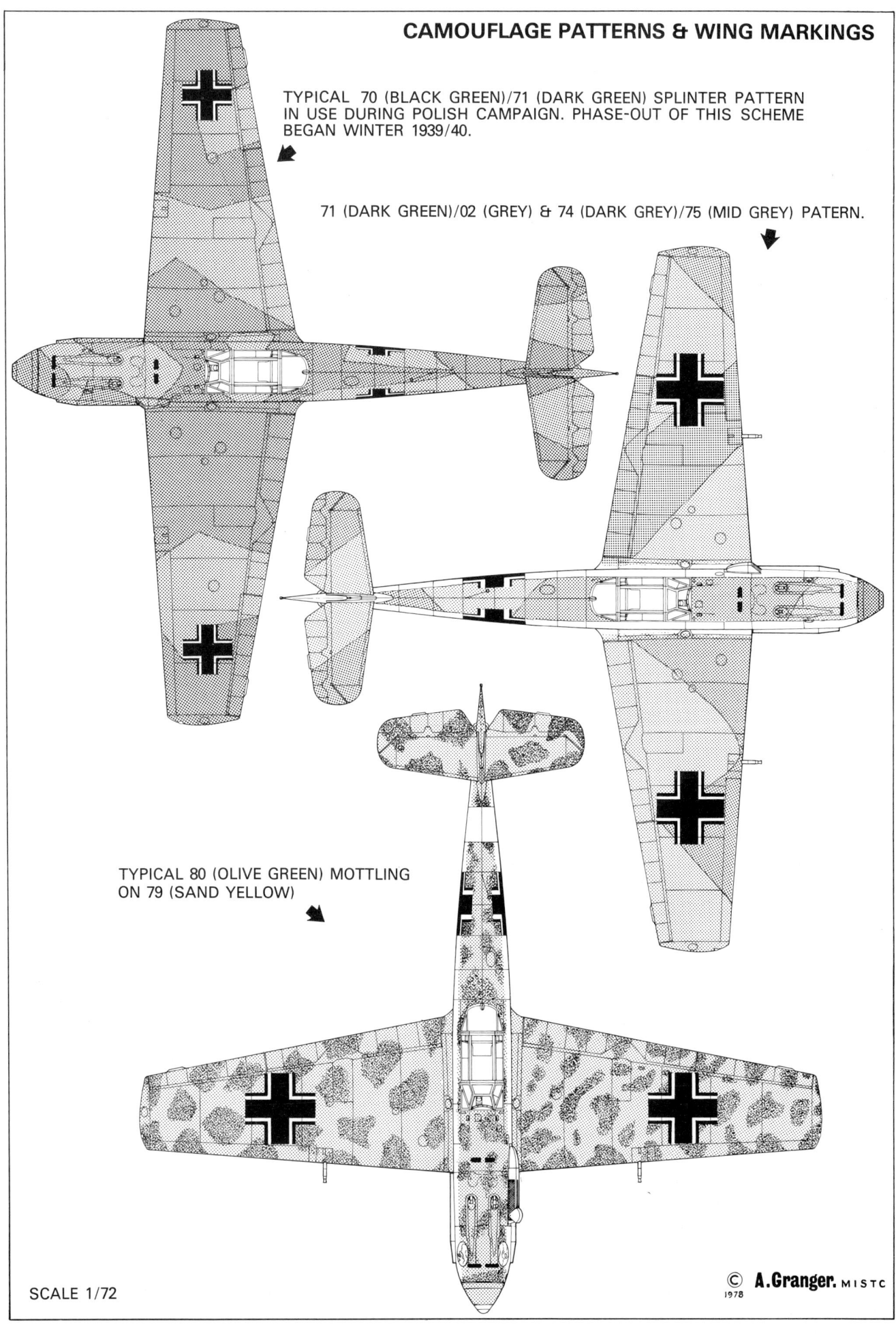

Fig. 14 *Adolf Galland as Commander of III/JG26 "Schlageter" in his yellow-nosed Me 109E-3, July 1940. Note telescopic sight.* [via Bruce Robertson.]   Fig. 15 *Galland in another Emil of JG26.*

was not reintroduced. Gone, too, were the MG FFs from the wings in favour of a pair of MG FFs, although the MG 17s above the cowling were retained. This particular model of the Messerschmitt 109 appears to have been the first one on which the entire production batch was fitted from the outset with the flat-topped hood with its associated armour plate behind the pilot's head, and although this was retrospectively introduced on earlier models many flew throughout the Battle of Britain without this modification.

During this same period reports began to appear stating that the guns were jamming after only two or three rounds had been fired, and this was eventually traced to the freezing up of the breech mechanisms at altitude. This problem could not be immediately dealt with at the factories and it is on record that experiments conducted at field level by JG26 in the heating of the guns ensured that the problem was obviated immediately in that unit at least.

Inevitably there followed the fighter-bomber version of the new model with the designation E-4/B. At the time this caused some excitement in the British press, which was conditioned to the seeming invincibility of the dive-bomber following the success of the Junkers Ju 87 in the earlier attacks on Poland and the Low

Fig. 16 *The Me 109E-7. Except for having provision for either a 300 litre drop tank or a bomb load varying from a single 50kg bomb to one of 250kg, this model was identical to the E-4N.*

Countries and under such headings as "Nazis Use Fighters As Dive-Bombers" reports that "They carried one medium bomb, they made one dive attack at 400 miles per hour and then they tried to escape" are typical of the period. In the event it is true that the bombs were delivered at quite a steep trajectory, and to aid in this a red line was frequently painted on the cockpit cover at 45 deg to the horizon so that when the pilot lined this up parallel the machine was presented at the correct diving angle. The modification for employment as bombers was carried out by means of the same "Rustsatz" kits as were issued for the E-1/B but this was not employed as a suffix to the designation as it was subsequently.

The radio equipment at this period consisted of a receiver/transmitter over the waveband between 2.5 and 3.7 megacycles. This was continuously tunable but could be locked on any desired frequency, although this could not be altered in the air. Power was supplied from a 24 volt lead accumulator of extremely compact design of roughly 7 amp hours potential, while the valves were the usual 5-pin type of the period.

A type of this Messerscmitt 109E-4 existed for use in tropical climates and was easily identified by the long filter forward of the supercharger air intake on the port side of the nose, while a night fighter fitted with a DB.601N engine was also introduced.

As with the Focke-Wulf 190, the Messerschmitt 109 was the subject of some "double rider" experiments with capsules of low frontal area faired into the upper surface of the wings. Although this idea was tried out primarily for employment as fuel tanks, on the 109 some attempts were made to incorporate a transparent fairing into the forward section in order that a parachutist could be carried in the prone position and dropped over hostile territory to act as a spy.

The redoubtable DB.601Aa engine was returned to in the reconnaissance version, the E-5, which was armed with only a pair of MG 17 guns for self-protection, while

Fig. 17 *Rumanian Air Force Me 109E-4 sporting on its cowling the exhortation "Hai Fetito!" (Up Little Girl!).*

Fig. 18 *Me 109E-3s for Yugoslavia.* [via Bruce Robertson.]

Fig. 19 *A 250kg bomb-carrying Me 109E-4B of III/JG1.*
Fig. 20 *Head-on view of an E-4B with a 250kg bomb slung from the ETC 500 belly rack.*

the DB.601N provided power for the otherwise similar E-6; this engine was also fitted to the E-7 which, except for the drop-tank, may be considered a re-engined E-4. The former was the first version to be fitted for the 300 litre tank and it was carried on a standard ETC 500 bomb rack originally conceived for the carriage of an SC 500 bomb or one of the PD 500 variety which was armour-piercing although, as far as is known, the Me 109 was never utilised in a tank-busting capacity.

For use in North Africa there followed the E-7/Trop, identified by its pointed spinner, although a number of E-7s with this feature and filters could be seen over the Eastern Front's northern sector in 1942. A low-level version was also developed with an increased weight mainly due to increased armour and designated the E-7/U2, while the E-7/Z had GM1 methanol-injection supercharging.

Another engine change brought about the DB.601E-engined E-8 reconnaissance fighter which was followed by the largely similar but lighter E-9 fitted with a drop-tank in the same manner as the E-7 already described and an RB50/30 camera.

In addition to the use of Messerschmitt single-seaters by the Spanish and Swiss Air Forces some Emils were employed by the Royal Rumanian Air Force which flew E-4s during 1942; these were finished in splinter camouflage above with light blue underparts, often with white spinners and entire yellow noses, wide rear fuselage band and bands under the wing tips. The large fuselage number was also in yellow and the rudder carried vertical stripes of the national colours of blue (leading), yellow and red in progressively larger bands. The yellow national crosses were blue-edged with wide white outlines and had in the centre a blue (centre) yellow and red roundel.

Employed on the southern sector of the Russian Front in similar finish were the E-7s of the Slovakian Air Force. With yellow under the noses, wing tips and in the form of a wide fuselage band, these machines had no rudder markings while the national insignia consisted of a mid-blue and white version of the Nazi cross with a large red disc superimposed in the centre.

The Augsburg product was the subject of several interesting experiments, but space only permits mention of two: one, carried out in the winter of 1942-43, involved an E-8 fitted with fixed skis for winter operations in the East; the other experiment involved the rigid attachment of an E-3 above a DFS 240 troop glider.

It is not generally realised that in an attempt to challenge British naval superiority, work was begun as early as 1939 on what may be described as a naval version of the Me 109 for operation from the aircraft carrier *Graf Zeppelin*; apart from fitting catapult

Fig. 21 *Neat line-up of Me 109E-4s in 1940.* Fig. 22 *E-4 W/Nr 1480 under armed guard after being crash-landed—reportedly due to damage from Lewis gun ground fire—near Marden, Kent, on 5 September 1940. Pilot was Oblt Franz von Werra, Adjutant of II/JG3.*

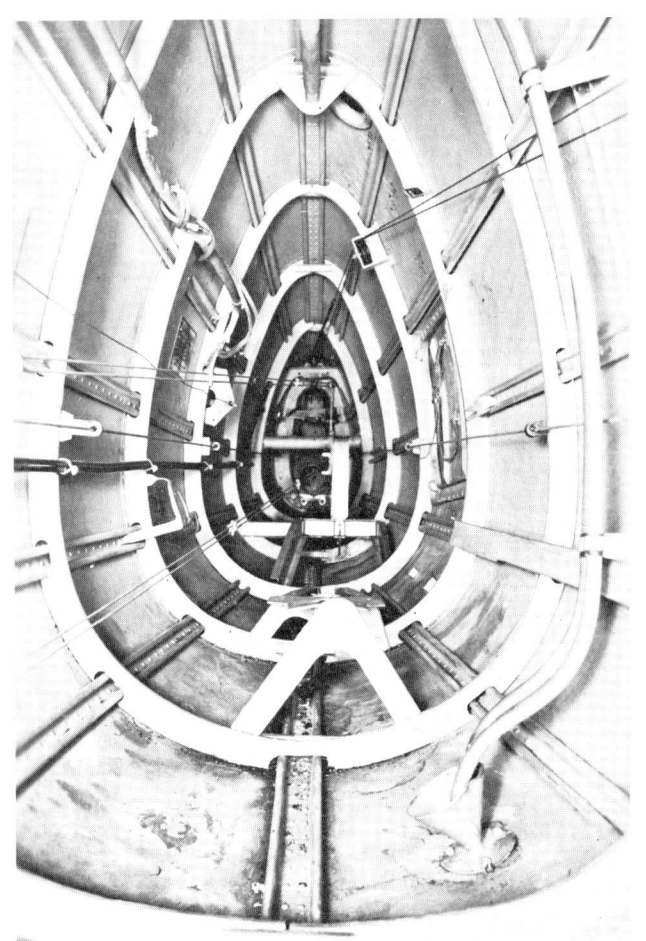

spools and an arrester hook, the main external difference consisted of an increase in wing span of 42½in (108cm) with an overall folded width for stowage between decks of 15ft ¾in (4.95m). The resultant fighter was redesignated the Me 109T, the suffix indicating Trager or Carrier. Detail design work was entrusted to Fieseler and ten E-1s were taken from the assembly line for conversion to pre-production Me 109T-0s with the DB.601A engines retained. These were fitted with retractable spoilers on the upper wing surfaces, to steepen the angle of glide, and a more robust undercarriage. However, by the time that trials were complete, work on the parent carrier had been all but halted so it was not until late 1940 that work was re-started with the assembly of 60 T-1s with the idea of employing them if work on the carrier was recommenced following the success of British shipborne fighters against the Italian Fleet at that time. In the event no such decision was made until some eighteen months later by which time the machines had been stripped of their naval equipment and issued as DB.601N-engined T-2s to I/JG77 (later known as I/JG5) for operations from small airfields. Later these were used for more experimental work from a dummy carrier deck near Bergen, but by May 1942 when work was again begun on the *Graf Zeppelin* it was considered that the Messerschmitt design had become too obsolete for trials to be continued.

Fig. 23 *A peep inside the rear fuselage of a Me 109 showing construction and cross-section.*   Fig. 24 *Close-up of the twin MG17 nose armament in von Werra's E-4, W/Nr 1480.*

SPECIFICATION
Messerschmitt Me 109E-3

Powerplant: One Daimler-Benz DB.601A twelve-cylinder inverted-vee liquid-cooled engine rated at 1,175hp for take off and 1,000hp at 12,140ft (3700m).

Dimensions: Span 32ft 4½in (9868mm); Length 28ft 4½in (8648mm); height (ground to vertical prop. tip datum horizontal) 12ft 0½in (3670mm); wing area 174.053sq ft (16.17m^2).

Performance: Maximum speed, fully loaded, 290mph (467km/h) at sea level; 307mph (494km/h) at 3,280ft (1000m). Maximum continuous cruise 300mph (483km/h) at 13,120ft (4000m). Maximum range 410 miles (660km). Rate of climb at 5,400lb (2450kg) 3,280ft/min (17.83m/sec). Time to 3,280ft (1000m) 1.1min. Service ceiling 34,450ft (10500m).

Armament: Two 20mm (Oerlikon MG FF cannon with 69rpg in wings and two fuselage-mounted 7.9mm Rheinmetall-Borsig MG 17 machine-guns with 1,000rpg (or 500rpg if MG FF/M installed). One engine-mounted 20mm MG FF/M cannon with 200 rounds mounted by some aircraft.

THE Me 109E SERIES

Type	Remarks
Bf 109E-0	Pre-production machines for evaluation.
E-1	First production model.
E-1/B	Fighter-bomber variant.
E-3	DB.601A replaced by DB.601Aa.
E-4	MG FF cannon in airscrew hub deleted.
E-4/B	Fighter-bomber variant.
E-4/N	Night fighter with DB.601N engine.
E-4/Trop.	Version for use in North Africa.
E-5	Reconnaissance version with two MG 17s.
E-6	As E-5 version but with DB.601N.
E-7	Provision for drop-tank.
E-7/Trop.	Filters and pointed spinner fitted.
E-7/U2	Armoured low-level model of E-7.
E-7/Z	GM1 supercharging.
E-8	Reconnaissance fighter with DB.601E.
E-9	Similar to E-8 with RB50/30 camera.
T-0	Ten pre-production carrier versions.
T-1	Sixty production shipboard fighters.
T-2	De-navalised version with DB.601N engine.

Fig. 25 *The DB601A engine installation in another Battle of Britain casualty. Nose guns have been removed.*

Fig. 26 *Me 109E-1 W/Nr 790 built in 1939 and allegedly flown in the Spanish Civil War. It was brought back from Spain in 1959 and is now displayed as shown—with the codes AJ + YH—in the Deutsches Museum, Munich.*

HAWKER HURRICANE I
By Philip J. R. Moyes

Fig. 1 *Brand new Hurricanes at Brooklands aerodrome before the war. They have the original Watts two-bladed wooden propellers but instead of the early "kidney" type exhaust stubs they have Rolls-Royce exhaust ejector stubs. Note the flight instrument venturis on the aircrafts' cockpit walls.*

Figs. 2 & 3 *First RAF squadron to receive the Hurricane was No 111—the famous "Treble One"—based at Northolt. In the ground view the nearest machine (again L1550) has the standard spearhead-shaped frame enclosing the unit badge on the fin.*

Wartime partner of the equally immortal Spitfire, the Hawker Hurricane was designed by Sidney (later Sir Sidney) Camm and his team, and stemmed from a design project begun in late 1933 for a monoplane development of the famous Fury biplane interceptor. Whereas the latter had, in its contemporary mark I form, a 525hp Rolls-Royce Kestrel engine, the new fighter, austerely referred to as the Fury Monoplane, was planned to have the 660hp Rolls-Royce Goshawk steam-cooled engine. However, when, in January 1934, Sidney Camm saw the initial design performance figures of the new Rolls-Royce PV.12, later to become the famous Merlin, the projected fighter was altered to take full advantage of this engine's greater power. With the change of engine, the fighter project was renamed the Interceptor Monoplane and, as design work progressed, an inwards-retracting undercarriage replaced the original fixed spatted undercarriage and the pilot's cockpit was enclosed by a transparent canopy.

Hawker received a contract for one prototype in February 1935, and on 6 November that year the new silver monoplane (K5083) made its first flight, from Brooklands, piloted by P.W.S. "George" Bulman, the company's chief test pilot. Unlike the Spitfire, which was of modern, light-metal stressed-skin monocoque construction, the Hurricane retained the time-honoured metal tube construction and fabric covering as used in Hawker aircraft since the late 1920s. The fuselage and tail unit were basically adaptations of the Fury's, and even the wings were fabric-covered over metal structures similar to those of the biplane. The fabric-covered wings were viewed with misgivings in some quarters and even before flight trials began, Hawker's project team initiated a design study into the provision of metal

Fig. 4 *The prototype Hurricane, K5083, which first flew on 6 November 1935.*

wings; however, these did not appear on production aircraft until 1939.

About three months after its first flight, the prototype went to the Aeroplane and Armament Experimental Establishment at Martlesham Heath, in Suffolk, for preliminary evaluation by the RAF, and although some teething troubles were experienced with the Merlin C engine, K5083 attained a maximum speed of 315mph

Fig. 5 *The first production Hurricane I, L1547. The film used caused the yellow outer rings of the roundels to become almost invisible.*

Fig. 6 *L1648, one of the first batch of 600 Hurricanes ordered, shows off its camouflage scheme for the photographer. In 1940 it served with No 85 Squadron.*

Fig. 7 *Hurricane Is of No 111 Squadron, sporting variegated unit numerals, arrive at Villacoublay, in France, in July 1938 to take part in a flypast over Paris as part of France's Bastille Day Celebrations.*

(523km/h) at 16,200ft (4938m) and climbed to 15,000ft (4572m) within 5.7 minutes of unstick. In March 1936 Hawker, confident that quantity production orders were imminent, began work on production drawings, and, simultaneously, draft schedules to cover an arbitrary potential output of 1,000 aircraft. Its confidence was rewarded on 3 June when a production order for 600 aircraft was received, and later that month the new fighter was officially named "Hurricane".

Production was delayed somewhat as a result of a decision to adopt the Merlin II engine instead of the Merlin I, and it was not until 12 October 1937 that the first Hurricane I (L1547) flew, at Brooklands. The first squadron to be equipped with what was in fact the RAF's first fighter capable of a top speed of over 300mph (483km/h) was No 111 at Northolt, beginning in December 1937. The CO, Sqn Ldr J. W. Gillan, made headline news in February 1938 by flying his Hurricane from Edinburgh to Northolt at an average speed of 408.75mph (657.8km/h)—albeit assisted by a strong tailwind. In April 1938 the first Hurricane with metal stressed wings flew, and in the months immediately preceding and following, the first Hurricanes ordered by Yugoslavia and Belgium flew. Other foreign customers who received small quantities of Hurricanes were Canada (where the type was also built under licence by the Canadian Car and Foundry Co), South Africa, Persia, Poland, Rumania and Turkey. Only one aircraft reached Poland before the Nazi invasion, and the invasion of Belgium put an end to the programme whereby Avions Fairey (Société Anonyme Belge) built the Hurricane under licence after only two aircraft had been completed.

At the outbreak of World War 2, RAF Fighter Command had 18 Hurricane squadrons as against nine full squadrons of Spitfires. In the early week of the war four squadrons went to France, Nos 1 and 73 Squadrons to serve alongside the Fairey Battle and Bristol Blenheim bombers of the Advanced Air Striking Force, and Nos 85 and 87 with the Air Component of the British Expeditionary Force. All the Hurricanes serving in France initially had the original two-bladed wooden propellers, and it was such a machine of No 1 Squadron which, on 30 October 1939, claimed the first enemy aircraft shot down on the Western Front; the pilot was Plt Off P. W. O. "Boy" Mould, and the victim a Dornier Do17 over Toul. In the following month, the very same Hurricane (L1842) had half its rudder and one elevator smashed off during an aerial collision with a French Air Force Morane fighter; the pilot, Sgt Clowes, managed to get the crippled Hurricane back to base but crashed on landing, albeit without injury to himself.

Hurricanes saw action with the Finnish Air Force during that first winter of World War 2—fighting the Russian Air Force for six weeks until the peace treaty ended hostilities on 12 March 1940.

In April the Germans invaded Norway, and on 26 May eighteen Hurricanes of No 46 Squadron were taken to Norway by the aircraft carrier HMS *Glorious*. The first three nosed over in the soft surface on arriving at

Fig. 8 *Aircraft of No 56 Squadron, North Weald, which, in April 1938, became the third squadron to receive the Hurricane. On 6 September 1939, two of its Hurricanes were accidentally shot down by Spitfires of No 74 Squadron, as related in the writer's monograph "Supermarine Spitfire Remembered" (VAP 1975).*

Fig. 9 *Hurricane I L1707 was displayed on Hawker Siddeley's stand at the 1938 Paris Air Show, afterwards being allotted to No 79 Squadron at Biggin Hill. Note triple ejector exhaust manifolds and ventral fin. Latter was introduced on production aircraft from March 1938 and at the same time the hitherto retractable tailwheel was fixed.*

Fig. 10 *Fine detail shot of L1791 with Hawker test pilot R. C. Reynell standing in the cockpit and Lord Nuffield looking on. This machine subsequently went to No 46 Squadron at Digby.*

Fig. 11 *Hurricanes of Treble One Squadron, carrying the pre-war codes "TM" (L1822 TM-R nearest), prepare for a practice scramble from their base at Northolt.*

Fig. 12 *A stannic "bomb" is used as a wind-indicator for Hurricanes of No 87 Squadron in France during the "phoney" war.*

their proposed base at Skaanland, so the remainder were diverted to Bardufoss to join No 263 Squadron's Gladiators. These two squadrons were the only RAF fighter units to take part in the ill-starred Norwegian campaign. On 7 and 8 June, most of No 46 Squadron's remaining Hurricanes returned safely to HMS *Glorious* to be evacuated, but only a few hours later the carrier was sunk by the *Scharnhorst* and *Gneisenau*; only two of the pilots survived, one of them being the squadron's CO.

During the Battle of France in May and June 1940, many RAF Hurricane squadrons besides those already mentioned saw action against the German invaders, including several home-based units. Losses were extremely heavy and in the final stages of the bitter campaign many Hurricanes, deprived as they were of spares or fuel, had to be destroyed on their airfields to prevent them from being captured. Even so, the Hurricanes had made their presence felt, as was seen, for instance, during the first seven days of the offensive in northern France when the Air Component's "Hurris" destroyed between 60 and 70 aircraft for a loss of 22 of their own in combat.

Meanwhile, the Merlin III and the earlier Merlin II had become the standard powerplants for the Hurricane I, and aircraft with these engines driving three-bladed metal variable-pitch propellers had begun to leave the production lines early in 1939.

When the Luftwaffe began its massive attacks against Britain's airfields and radar stations as well as her

Fig. 13 *Some of No 73 Squadron's Hurricane Is over France during the "phoney" war. Formation includes P2569 D, P2575 J and N2358 Z. With No 1 Squadron No 73 formed 67 Wing, Advanced Air Striking Force, in France and both squadrons deleted their code letters but carried red, white and blue rudder stripes for the benefit of trigger-happy French fighter pilots.*

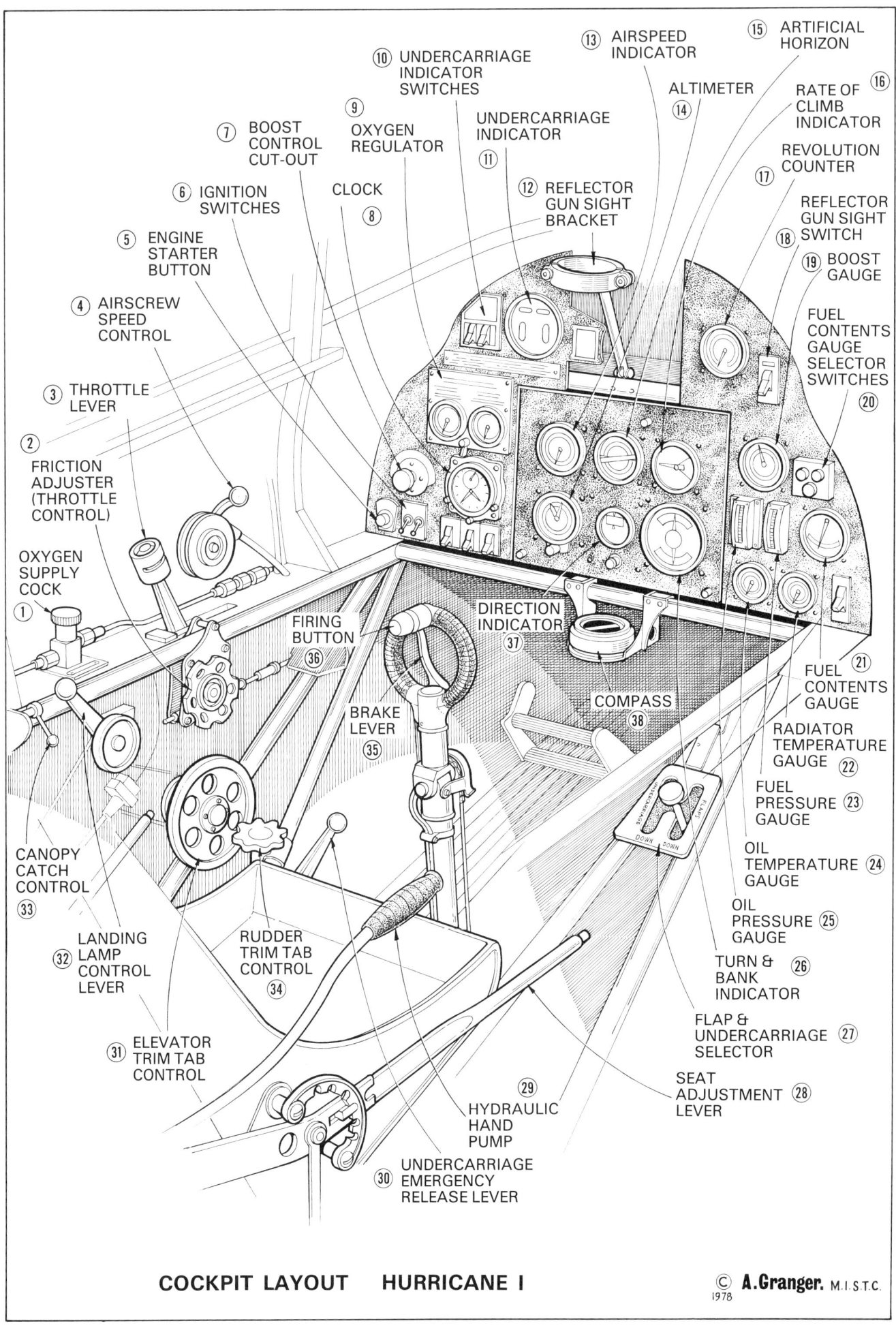

COCKPIT LAYOUT   HURRICANE I

HURRICANE I
(Flg Off CLOWES)
NO 1 SQUADRON
WITTERING
OCTOBER 1940.

SCALE 1/72

RESEARCH: A. GRANGER © 1978
ARTWORK: ROY MILLS

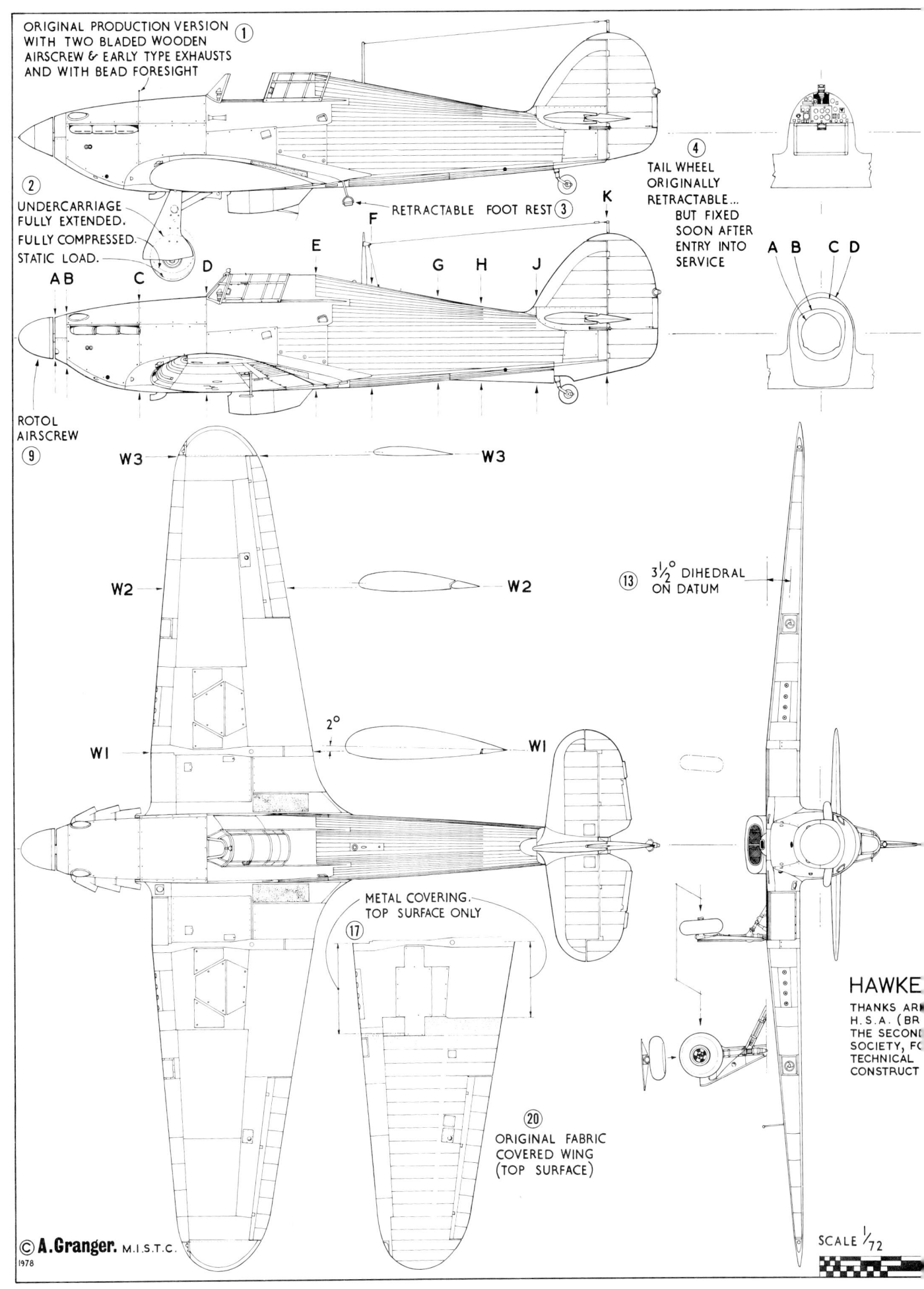

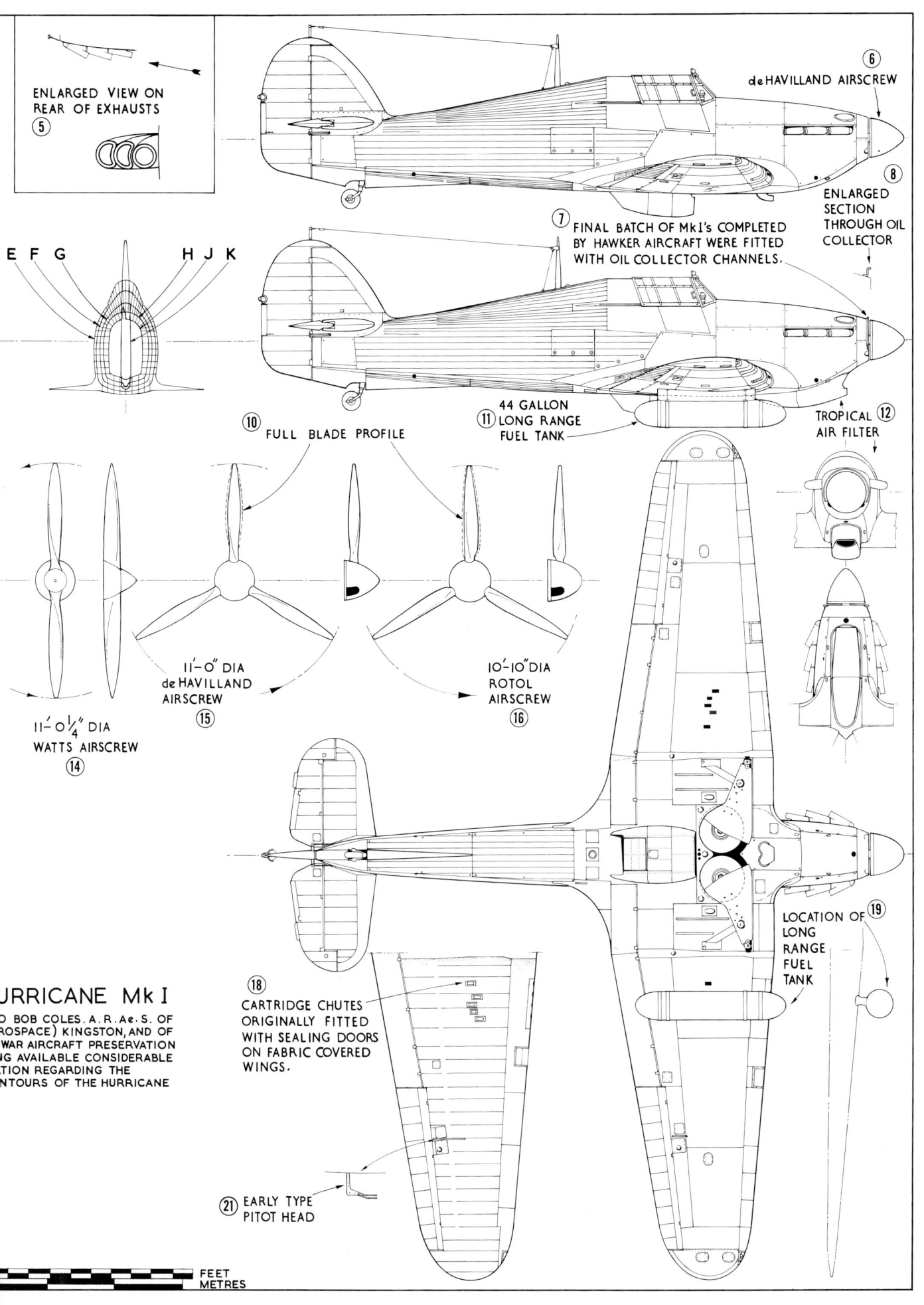

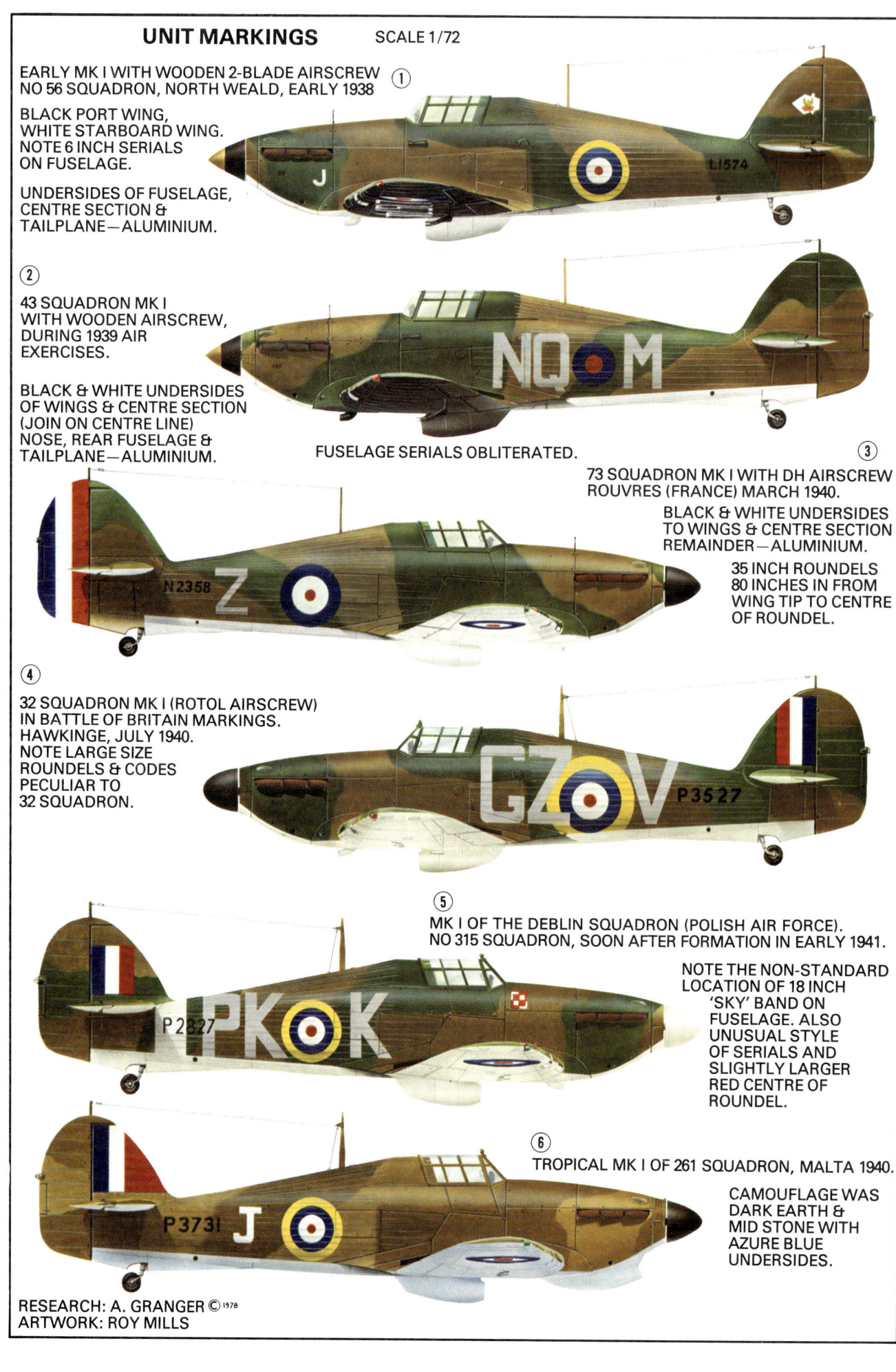

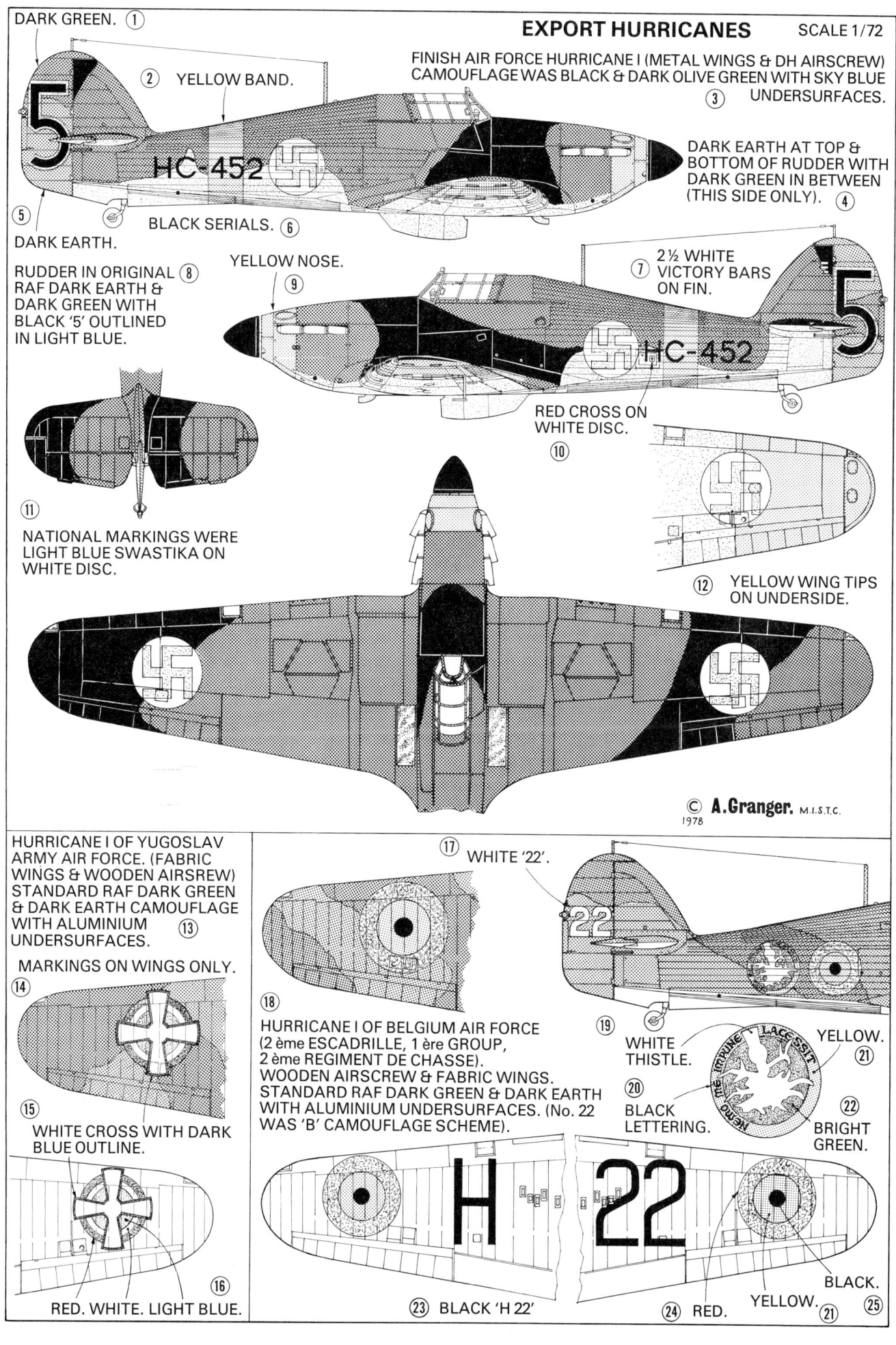

Figs. 14 & 15 *Hurricanes of No 85 Squadron above cloud-covered England towards the end of the Battle of Britain. Note how the "Sky" undersurface colour has been extended up the side of the nose of P3408 VY-K.*

Channel convoys on 12 August 1940, No 11 Group's thirteen squadrons of Hurricanes, six squadrons of Spitfires and two of Blenheim IFs based in south-eastern England had to take the full shock of it. The tactics planned by Fighter Command's famous AOC-in-C, Sir Hugh Dowding, and Air Vice-Marshal K. R. Park who commanded No 11 Group, were for the Hurricanes to intercept the high-flying bombers while the Spitfires tackled the higher-flying fighter escort. This was a logical scheme, for at their rated altitude of around 15,000ft (4572m) the Hurricanes were a match for any Luftwaffe aircraft of that time, and the attacking bombers rarely flew above 17,000ft (5182m). The engines of the Spitfires, on the other hand, were rated for optimum performance at 18,000ft (5486m). In practice, these tactics could not always be followed, for various reasons, including the unforseen one that some of the combats would take place well above 20,000ft (6100m) where the Me 109 could out-perform even the Spitfire I.

The success of the Hurricane in the Battle of Britain has been well recorded elsewhere, but it is worth mentioning here that this redoubtable fighter destroyed more enemy aircraft during the battle than the combined total of all other defences—aircraft and anti-aircraft guns. However, in fairness to the Hurricane's fighting partner, the Spitfire, it is only right to point out that throughout the battle the Hurricane squadrons were more numerous. Incidentally, the first Hurricane IIAs began to reach the squadrons from 4 September 1940, but of course the Mk I constituted the bulk of Fighter Command's Hurricane element for several more months.

Recalling some of the critical days of late August and early September, one Hurricane pilot afterwards wrote: "These were hectic days and the only time we saw the pilots of other squadrons was when we met in the mess during the evening after long hours in the air or at readiness. But often, just as one was becoming friendly with a pilot he would get shot down and we would see him no more.

"When we met the Hun we 'mixed it' well and truly. Usually we waited for a moment or two until we were in a favourable position before making the first attack. After that it was every man for himself. There were always more Huns than ourselves. We whirled around taking squirts at as many machines as possible; sometimes when they burst into flames, the crew baled out or we were lucky enough to see where they hit the ground; then we were able to claim victories.

"After the battle we would rush back to our base, tell our stories to the Intelligence Officer, then ask him to get through to the hospitals to see if a friend had got any before he was shot down. I remember the German bombers, flying in a tight formation, crossing the

Fig. 16 *A Merlin III-powered Hurricane I from Gloster's third main production batch.*

English coast and thinking to myself that they were about to drop their bombs in England, and that I was in a Hurricane and could stop them. I dived to attack and laughed as they broke formation; the crew of one Heinkel baled out; the others sprayed me with tracer—many bullet holes in my plane that time, but within an hour the ground crews had fixed it and my Hurricane was ready for action again.

"Our life had a devil-may-care sort of happiness but often as we lay in the sun near our machines, waiting at readiness, there were moments of great beauty; somehow the colours in the field seemed brightest and the sky the deepest blue just before taking off to meet another raid. At dusk everything became peaceful. We were all tired but happy at the thought of another day's work accomplished. Our Hurricanes stood silhouetted against the sky looking strong and confident, the growing darkness hiding the patched-up paintwork. The following morning we awoke to the roar of engines being tested for another day's work."

When the Luftwaffe attacked Gosport on 16 August 1940, one Hurricane pilot, Flt Lt J. B. Nicolson of

Figs. 17 & 18 *Hurricane Is of No 601 Squadron are refuelled and re-armed at Tangmere in 1940. Pilot at left in view below is Max Aitken.*

Fig. 19 Offering more close detail for the modeller is this evocative study depicting an aircraft of No 32 Squadron during the Battle of Britain.

Fig. 20 *A winged Popeye insignia decorated this machine of No 17 Squadron photographed at Debden in September 1940.*

No 249 Squadron, based at Boscombe Down, won what proved to be the only Victoria Cross awarded to RAF Fighter Command. The section of three Hurricanes he was leading was just about to attack some Me 110s when it was bounced from astern by Me 109s and all three Hurricanes were hit. Four cannon shells struck Nicolson's machine. One of them tore through the cockpit hood and sent splinters into his left eye, almost severing the eyelid and temporarily blinding him in that eye. Another shell struck and exploded the reserve fuel tank behind the instrument panel, which set the Hurricane on fire. The third crashed into the cockpit and tore away his right trouser leg. The last hit his left foot and wounded his heel.

Nicolson swerved and dived to avoid further shells, and finding a 110 ahead of him he dived after it at about 400mph (644km/h). Getting the 110 in his gun sight, he pressed the firing button and, as he did so, he saw his right thumb blistering in the heat. He also saw his left hand, holding open the throttle, blistered in the flames. The instrument panel was shattered and "dripping like treacle".

"Curiously enough", he later recalled, "although the heat inside must have been intense, in the excitement I did not feel much pain. In fact, I remember watching the skin being burnt off my left hand. All I was concerned about was keeping the throttle open to get my first Hun."

The Messerschmitt weaved from side to side to avoid the hail of bullets from Nicolson's now blazing Hurricane. In the cockpit the heat was so great that Nicolson put his feet on the seat beneath his parachute while he continued to fight until the 110 disappeared in a steep dive. Nicolson then tried to bale out of his aircraft and struck his head on the framework of the hood —which was all that remained of it. He threw it back and tried to jump again. This time he realised that he had not unpinned the harness straps holding him into his seat. One broke, doubtlessly burnt. He undid the other and at last jumped out, only to be wounded in the buttocks by shotgun pellets fired at him by an overzealous Local Defence Volunteer as he floated down by parachute. In addition to the VC, Nicolson won the DFC later in the war, and was posted missing, believed killed after a Liberator bomber crash in the Far East in May 1945.

RAF Hurricane Is went into action in the Middle East and Mediterranean in 1940; and in the following year, when the Germans invaded Yugoslavia, about 40 Hurricanes were serving with the Royal Yugoslav Air Force. By that time Hurricane production under licence had begun in Belgrade, but the enemy's occupation of the country brought it to a close. The Germans swept through Yugoslavia and on into Greece where Hurricanes and Gladiators of Nos 33, 80 and 112 Squadrons made a courageous but utterly hopeless stand against both the German and also the Italian air forces. The British fighters were soon overwhelmed and the pilots who survived withdrew first to Crete and finally back to Egypt from whence they had first come. By now,

Fig. 21 *More good detail here in a view of L1562, sixteenth production Hurricane, which was used for flight trials of the de Havilland-Hamilton two-pitch three-blade metal propeller in 1938.*

Fig. 22 This fine air-to-air study of L1683 shows to advantage how the thrust tolerance on the propeller shafts of early Hurricanes resulted in a considerable "gap" between propeller hub and nose cowling. Later machines incorporated a flanged plate on their noses which hid the gap.

though, the Luftwaffe was operating in North Africa and the outmoded Hurricane Is were greatly handicapped by the bulky tropical sand filters, fitted over the carburettor air intakes, which reduced their top speed to little over 300mph (483km/h).

Hurricane Is fought with RAF squadrons in the Far East during the early stages of the campaign there—in Singapore, the Dutch East Indies, Burma and Ceylon. Following the fall of Singapore, 24 Mk Is were handed over to the Dutch in Java and operated by the Dutch Java Air Force during February and March 1942. One Squadron defended Batavia, whilst other machines attacked invading Japanese seaborne forces at Bantam Bay and Kretan. In just over two weeks, 30 enemy aircraft were destroyed by the Dutch pilots at a cost of 18 Hurricanes.

A total of 3,924 Hurricane Is were eventually completed in the UK—1,924 of them by Hawker and the rest by Gloster. The first Canadian-built Hurricane I (P5170) flew on 9 January 1940 and machines from the Dominion began to arrive in the UK in time to fight the the Battle of Britain. All told, 160 Canadian Hurricane Is were built, being eventually redesignated Mk X as they differed from the British-built aircraft in having American Packard-built Merlin 28 engines.

SPECIFICATION

Powerplant: (late production aircraft) one 1,030hp Rolls-Royce Merlin III twelve-cylinder liquid-cooled vee engine.
Dimensions: Span 40ft 0in (12192mm); length 31ft. 5in (9576mm); height with airscrew vertical—flying attitude 12ft 0½in/12ft 2in (3670mm/3708mm) according to airscrew fitted; wing area (gross) 258sq ft (24m²).
Weights: (empty) 4,982lb (2260kg); (loaded) 6,447lb (2924kg).
Performance: Max speed 254mph (409km/h) at sea level; rate of climb at 11,000ft (3353m) 2,420ft/min (738m/min); max range with 20 min reserves 425mls (684km); service ceiling 34,200ft (10970m).
Armament: Eight .303in (7,7mm) Browning machine-guns with 334 rounds per gun.

Fig. 23 *Another view of L1562 with its as yet unpainted DH-Hamilton two-pitch three-bladed propeller. Test pilot is P. G. Lucas.*

REPUBLIC P-47D THUNDERBOLT

By John B. Rabbets

Fig. 1   *P-47D-11-RE 42-75587 "Li'l Sunshine" of the 379th Fighter Squadron, 362nd Fighter Group, Ninth AF, on a sortie from Wormingford, near Colchester, early in 1944.*

Fig. 2 *Head-on view of the Republic Thunderbolt emphasizes the deep oval section fuselage, large radial engine driving a 12ft (3657.60mm) diameter propeller, projecting oil cooler shutters and four degree dihedral wing.* [Republic via R. A. Freeman]

The P-47D was the most numerous version of the most produced American fighter plane of World War 2, the Republic Thunderbolt. Of more than fifteen and a half thousand P-47s made between May 1941 and December 1945, 12,602 were D-models constructed at Farmingdale, Long Island and Evansville, Indiana from December 1942 onwards. The 354 P-47Gs built by Curtiss Wright at Buffalo were essentially D-models too.

The Thunderbolt served with Free French, Soviet, Brazilian and Mexican forces in addition to the USAAF and Royal Air Force. Four hundred and forty-six P-47Ds went to the Free French, and some remained in service until 1954 against Algerian rebels. One hundred and ninety-six P-47D-22-RE and P-47D-27-RE Thunderbolts reached Soviet Russia out of 203 originally despatched via Persia. The 1st Fighter Squadron of the Brazilian Air Force operated some of the eighty-eight Brazilian Thunderbolts as a fourth squadron with the 350th Fighter Group of the US 12th Air Force in Italy from October 1944. Early in 1945, with its training complete, the 201st Fighter Squadron of the Mexican Air Force joined US forces in the South West Pacific with 25 P-47Ds. Japan surrendered before it became operational. The 830 P-47Ds lend leased to the Royal Air Force operated exclusively with South-East Asia Command.

Progressive development was the hallmark of the D-model. The early D closely resembled the C in most respects, whereas the late D-model was comparable with the M in power and performance. It had the cut-down, more streamlined rear fuselage and bubble canopy which gave the pilot vital all-round vision. The Thunderbolt had great consistency and harmony of form. The outline of the partly elliptical planform wing with its straight leading edge and curved trailing edge, was echoed in the similar shape of the vertical and horizontal tail surfaces, and the deep curved fuselage of generous proportions harmonised with the flying surfaces.

Of the USAAF's five main wartime single-seat fighter types, only the P-47 had a radial engine. This was the large, extremely reliable two-row Pratt & Whitney R-2800 motor installed in a deep, low-drag cowling. Positioned midway along the fuselage, the cockpit took up less than half the available depth. The whole lower part of the oval section monocoque structure was occupied by ducting which conveyed air and exhaust gases under the wing to and from the heavy supercharger installed just forward of the retractable tailwheel. Russian-born designer Alexander Kartveli first positioned and balanced the large engine and bulky turbo-supercharger and then completed the design of the rest of the fuselage. The tremendous power of the engine was absorbed by an unprecedented 12ft 2in (3709mm) diameter Curtiss Electric or 13ft 1⅞in (4010mm) diameter Hamilton Standard Hydromatic four-bladed propeller. The four degree dihedral wing of Republic S3 section was built around two main spars and three auxiliary ones which supported ailerons, flaps and undercarriage. Like the fuselage it was covered with flush-riveted Alclad stressed skin. Giving the necessary ground clearance, the long inward-retracting undercarriage units had to shrink nine inches (229mm) during retraction to allow the eight .50in (12.7mm) calibre Browning guns to be accommodated within the deeper midspan portion of the wing whilst giving adequate ammunition storage towards the tips. The guns were staggered in their bays to simplify ammunition feed.

Each gun weighed 65lb (29.5kg) and fired a mixture of incendiary, tracer and armour-piercing ammunition at 800 rounds per minute. Bombs and droptanks could be carried on streamlined underwing pylons located at the highly stressed aileron/flap joint or slung from the belly shackles despite the relatively restricted ground clearance. By July 1944, the Thunderbolt was considered the best ground-strafing machine available to the Allies. Moreover, it could absorb far more punishment that any other fighter type and yet bring its pilot safely back to base. Two 500lb (227kg) bombs on the wing racks were the most common load for ground attack. Fragmentation bombs were carried in clusters and P-47Ds occasionally carried 250lb (113.5kg) GP bombs in threes on each wing rack. From the P-47D-20-RE model, an offensive load of 2,500lb (1134kg) could be carried.

On 17 August, Lt Colonel David Schilling flew a P-47D with three-tube 4.5in (115mm) rockets under each wing on a mission against railway rolling stock, but the weapon was never popular. The P-47D-28-RA belonging to Colonel Frederick C Gray, CO of the 78th Fighter Group, for a time had a 20mm cannon slung from each wing rack. Vibration problems terminated this experiment.

Republic P-47Cs first arrived as deck cargo to join the US Eighth Air Force in England on 20 December 1942. After reassembly at depots such as Burtonwood, they were issued to the 4th, 56th and 78th Fighter Groups in January 1943, and to the 495th Fighter Training Group at Atcham. P-47Ds arrived in April 1943 and rapidly supplanted the C-models.

Enormous mercantile losses made shipping fighter planes across the Atlantic a risky business. So in August 1943, ten P-47Ds were experimentally air delivered to the Eighth Air Force via the North Atlantic ferry route. One of the pilots was Captain Barry M Goldwater, later famous as a Senator from Arizona. He flew a P-47D-5-RE, 42-8550, called "Peggy-G", with nine other Thunderbolts in stages from Farmingdale via Presque Isle, Goose Bay, Blueie West One (Greenland) and Meeks Field (near Reykjavik, Iceland) to Prestwick. For the flight each Thunderbolt carried two underwing 165 gallon (625 litre) tanks and a homing radio compass with D/F loop behind the cockpit. Each of the two flights of five Thunderbolts was led by a B-24E whilst ahead was a C-87 acting as overall flight leader. The stages were flown at ground speeds of 180-190mph (290-306km/h). Neither Republic nor Fighter Command nor Ferry Command liked the long over-water ferrying of single-engined single-seat fighters, and the operation was not repeated.

The 353rd Fighter Group joined the Eighth Air Force with Thunderbolts in August, the 352nd and 355th in September, the 356th in October and the 358th and 359th

Fig. 3 *In this view a P-47D-11-RE, 42-75568, clearly portrays the elliptical planform wing, pointed windscreen, framed sliding cockpit canopy and sharp rear fuselage spine. The wing guns and pointed hub of the Curtiss Electric propeller stand out sharply.* (Republic via R. A. Freeman]

Fig. 4 *The characteristic wide track undercarriage common to all models of the Thunderbolt is clearly shown in this ground shot of a P-47C. Also visible are the distinctive elliptical engine cowling and starboard wing root cockpit air conditioning intake.* [Republic via R. A. Freeman]

in December 1943. in January 1944 the 361st FG became the last Thunderbolt group to join the Eighth which had reached a peak strength of ten Thunderbolt groups. The 358th Fighter Group did not remain long with the Eighth. On 1 February 1944 it went to the Ninth Air Force in exchange for the 357th equipped with Mustangs. The majority of future long-ranging Mustang groups served with the Eighth and new Thunderbolt groups were allocated to the Ninth Air Force, reforming in England as a tactical air force for the impending invasion of Europe.

Before the Merlin-powered Mustangs finally fulfilled the ideal of escorts able to accompany the heavies all the way to the target, Eighth Air Force Thunderbolts had gained a degree of air superiority over the Luftwaffe. The first confirmed Eighth Air Force Thunderbolt victory over a German jet fighter had been on 29 August 1944, when Major Joseph C Myers and Lt Manford O Croy shared the destruction of an Me262 near Brussels during and evening mission by the 78th Fighter Group. When the 78th FG finally relinquished its three squadrons of P-47Ds in December 1944, and re-equipped with P-51Ds, only the 56th FG of the Eighth continued to operate the Thunderbolt.

In the two years between the first of 447 missions on 13 April 1943 and the last on 21 April 1945, the 56th destroyed 674½ enemy aircraft in the air—more than any other Eighth AF fighter group—and 311 on the ground, for a loss of 128 Thunderbolts. Among its 62 aces—more than any other USAAF group—were Colonel Francis S Gabreski and Captain Robert S Johnson each with 28 air victories. Close behind came Colonel David C Schilling and Major Frederick J Christensen both with 22½, Major Walker H Mahurin with 19¾ and the legendary Colonel Hubert A Zemke with 17 air victories whilst flying P-47s with the 56th. Half of Bob Johnson's victories were Fw190s, Dave Schilling shot down three Bf109s and two Fw190s on 23 December 1944 and Frederick Christensen was the first Eighth AF pilot to shoot down six enemy aircraft on one mission on 7 July 1944. Gerald W Johnson who became the group's first ace on 18 August 1943 was shot down on his 88th mission on 27 March 1944 with a score of 18 air victories.

In October 1943, it was known that the Mustang would be coming to VIIIth Fighter Command. Colonel Zemke, then group commander of the 56th, put in a bid to get the P-51 for the group. This did not materialise, and the unit continued to use the Thunderbolt until the end of the war. The P-47Ms which arrived in England in early 1945 not only offered the group a superior performance to the late D-models it had been flying, but with a top speed of 465mph (749km/h) at 32,000ft (9753m) the P-47M was faster than the Mustang. By the end of the war, the 56th FG had claimed seven Me262s and two Ar234s. With one exception they were downed in March and April 1945, when the group had been largely re-equipped with the P-47M.

Faced by nearly one thousand VIIIth Fighter Command fighters, German Bf109s and Fw190s preferred to molest unaccompanied or weakly escorted VIIIth Bomber Command heavy bombers in early 1944, and avoid battle with the P-38s and P-47s. Yet the alarming Luftwaffe fighter numbers had to be destroyed before D-Day, set for early June 1944. Colonel Glenn E Duncan, CO of the 353rd Fighter Group therefore suggested to Major General William E Kepner, Commanding General of VIIIth Fighter Command, that sixteen volunteers from four different P-47 groups be given specially intensive training in the art of ground

Fig. 5 *New P-47D-22-REs in natural metal finish at Farmingdale including 42-26077 in the foreground.* [Republic via R. A. Freeman]

Fig. 6 *More P-47D-22-REs at Farmingdale including 42-25974, 42-26065 and 42-26079. Blunt hub of the Hamilton Standard propeller and underwing stores pylons show in this photograph.* [Republic via R. A. Freeman]

Figs. 7 & 8 *Close-ups of a Republic P-47D-20-RA fitted with three launcher tube clusters for 4.5in (11.43mm) folding fin rockets under each wing.* [Republic via R. A. Freeman]

strafing. Under Duncan as CO, the unit, irreverently called "Bill's Buzz Boys", comprised four flights from the 353rd, 355th, 359th and 361st Fighter Groups. The P-47Ds they flew all had paddle-bladed propellers, to ensure good low altitude performance. In eight missions beginning 26th March 1944, the unit despatched 83 effective sorties. Three P-47s and two pilots were lost and 13 Thunderbolts suffered damage. But the unit claimed 14 enemy aircraft destroyed, six probably destroyed, 14 damaged on the ground, and one probably destroyed in the air. Seventeen locomotives, a boat and a hangar were accounted for and nine flak towers strafed. Then the unit disbanded and the flights returned to their home bases.

Bill's Buzz Boys changed the Luftwaffe's policy of conserving planes by keeping them on the ground except when they could be used to fullest advantage. From now on the German Air Force would either have to fight in the air where it could be mastered or be strafed and destroyed on the ground by roving Allied fighters.

The 5th Emergency Rescue Squadron was formed in the Eighth Air Force in early May 1944. Based first at Boxted and then at Halesworth it supplemented Royal Air Force Air-Sea Rescue services. It was equipped with war-weary P-47Ds from every P-47 group in VIIIth Fighter Command. Initially they each carried two 108 gallon (409 litre) external wing tanks, a container for two British M-type dinghies under the belly and four smoke marker bombs on racks under each wing behind the wheel wells. The useful load was soon revised to provide for a 150 gallon (568 litre) belly tank, an M-type dinghy pack on each underwing rack and four smoke markers aft of the belly tank.

Early in the morning of 13 June 1944, the first of 7,547 V1 flying bombs launched against England, pulsed its way towards London. On 30 June the first one to be shot down by an Eighth Air Force pilot was downed by Lieutenant J Tucker, a member of the 5th ERS flying a war-weary P-47D.

In its tactical operations, the Ninth Air Force showed what a superb ground attack machine the P-47 was, and how much punishment it could absorb. With fifteen groups of Thunderbolts, it was the largest user of the type. From February 1944 until VE Day, Thunderbolts were sometime or continuously the equipment of the 36th 48th, 50th 354th, 358th, 362nd, 365th, 366th, 367th, 368th, 371st, 373rd, 404th, 405th and 406th Fighter Groups. Except for the Distinguished Unit Citation awarded to the 365th FG for an aerial duel on 21 October 1944 over the Bonn-Düsseldorf area, all the 21 DUCs won by the groups were for ground support and strafing attacks against airfields and enemy ground forces on the Continent after D-Day.

A typical unit was the 404th Fighter Group which arrived in the United Kingdom on 3 April 1944. Its first operational sortie with 48 Thunderbolts took place on 1 May 1944 under Lt Col C W McColpin, an Eagle Squadron veteran. The group was engaged on operations over the Normandy bridgehead until 7 June. Subsequently it moved to the Continent to fly escort missions. Its 506th Squadron had razorback D-models, whilst its other two squadrons had bubble-canopied Ds. The 404th's last operations were flown from Kelz Fritzlar in Germany. In spring 1945, a black "Thunderbird" motif became standard on the nose cowls of the group's P-47Ds.

The Twelfth Air Force began missions with Operation Torch, the landings in North West Africa, on 8 November 1942. But its six fighter groups eventually equipped with Thunderbolts had to wait another year for the first P-47s. The 57th Fighter Group had its P-47s for the longest period, November 1943 to May 1945, and the early machines came in olive drab finish. Like the 57th, the 79th and 350th Fighter Groups moved to Italy with their P-47s in September 1944. The other three Twelfth Air Force P-47 groups, the 27th, 86th and 324th FGs were transferred from the 12th AF to the 1st Tactical Air Force, the 324th in November 1944 and the 27th and 86th in February 1945.

Lt Raymond L Knight of the 350th FG was awarded the Congressional Medal of Honor for missions against heavily defended airfields in Northern Italy when he personally accounted for 20 aircraft in two days. His Thunderbolt crashed from flak damage in the Apennines on 25 April 1945 and Lt Knight died as he vainly tried to regain his base. His Medal of Honor was the only one awarded to an MTO pilot and to a P-47 pilot in Europe.

On 1 November 1943, a new strategic air force, the Fifteenth, began to be built up to use the Foggia airfield complex of southern Italy. It was predicted that winter weather there would allow the Fifteenth's bombers to make twice as many attacks on German industry compared with those of the UK-based Eighth. Targets in southern Europe and the Balkans could also be reached and German defences further stretched and weakened.

Fig. 9 *42-23278, a P-47D-15-RA restored by Republic and flown at the 25th Paris Salon at Le Bourget in 1963. Bearing the civil registration N5087V and the codes HV-P it was piloted by Glenn C. Bach, an ex 359th Fighter Group captain of WW2.* (Republic via R. A. Freeman]

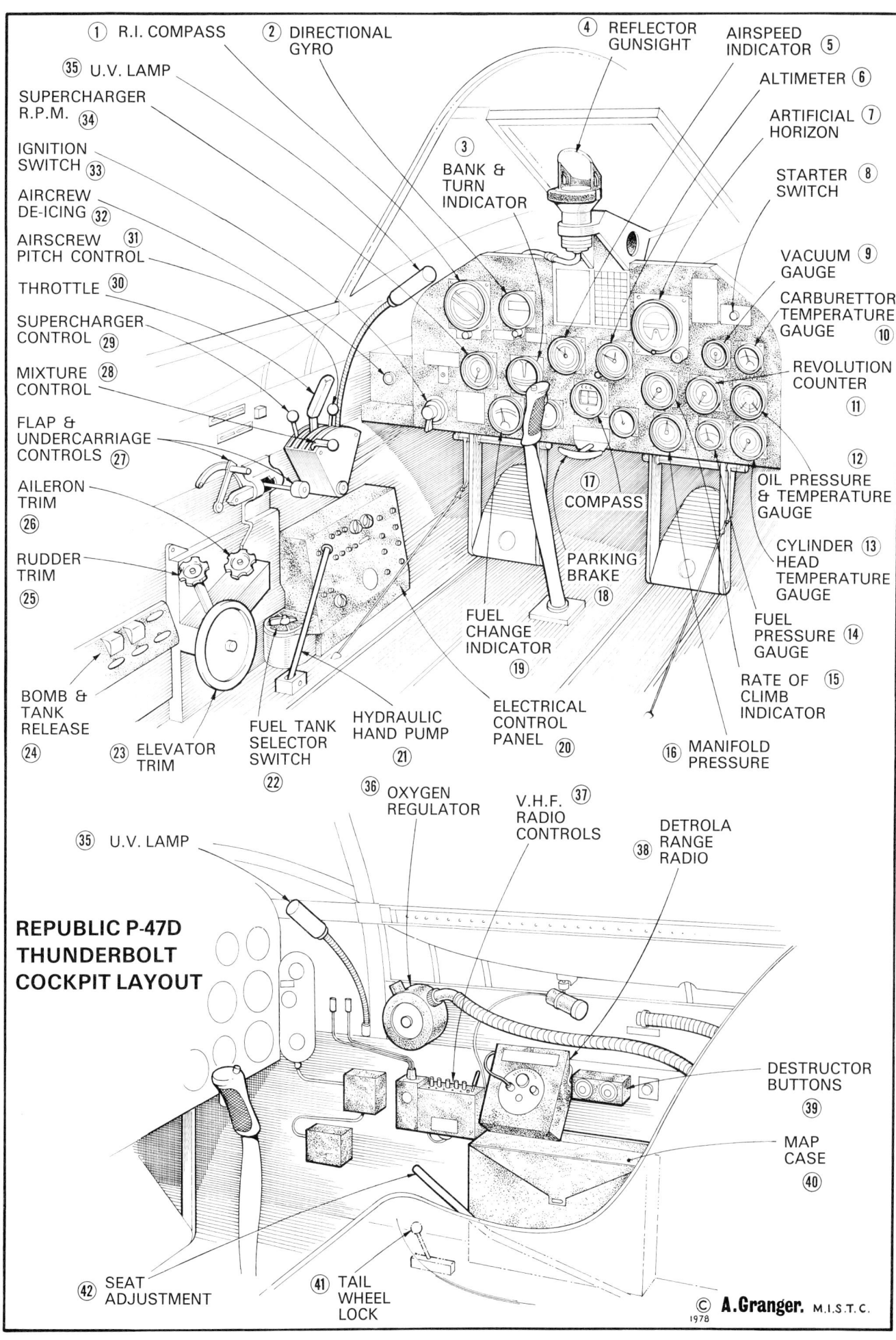

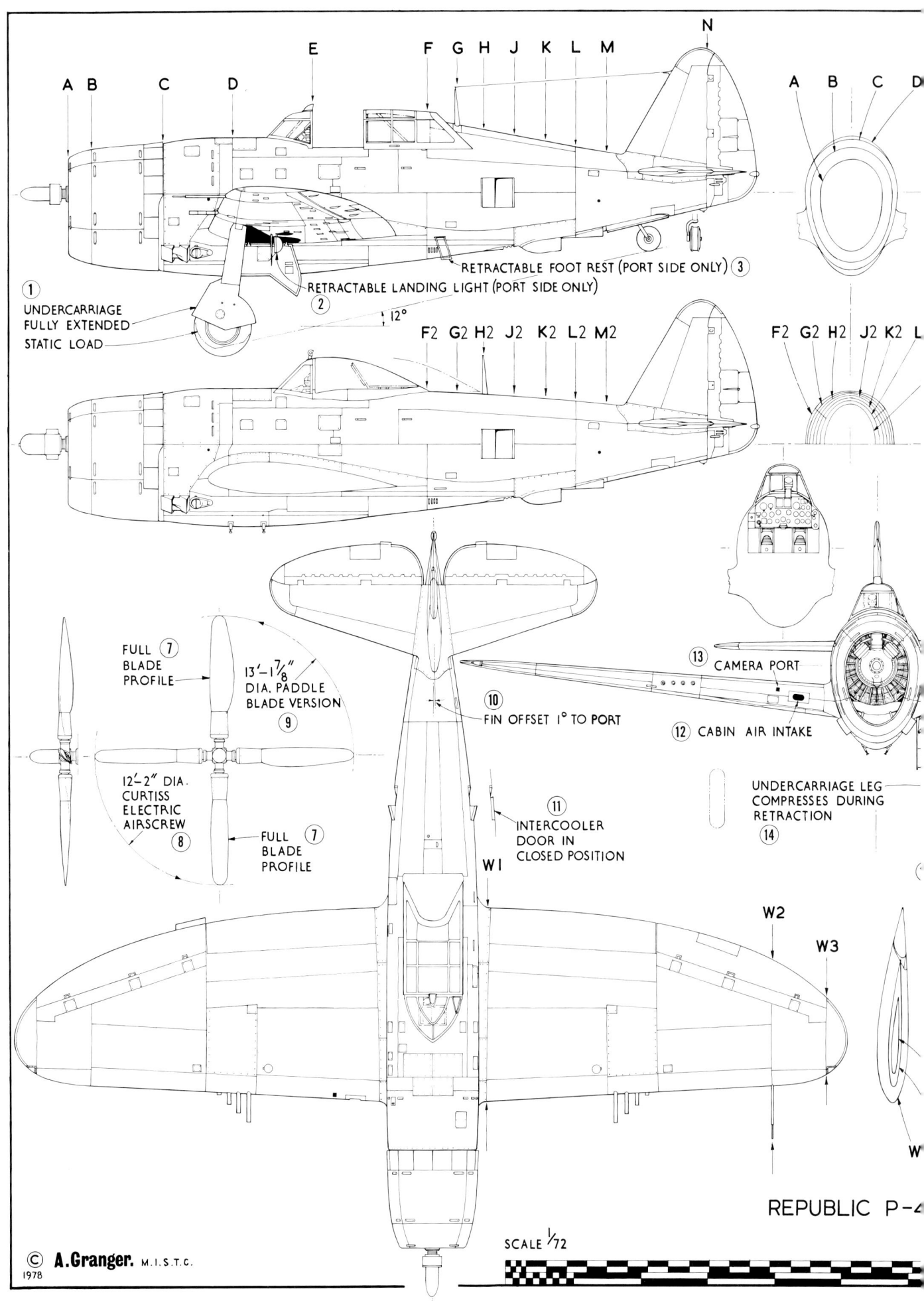

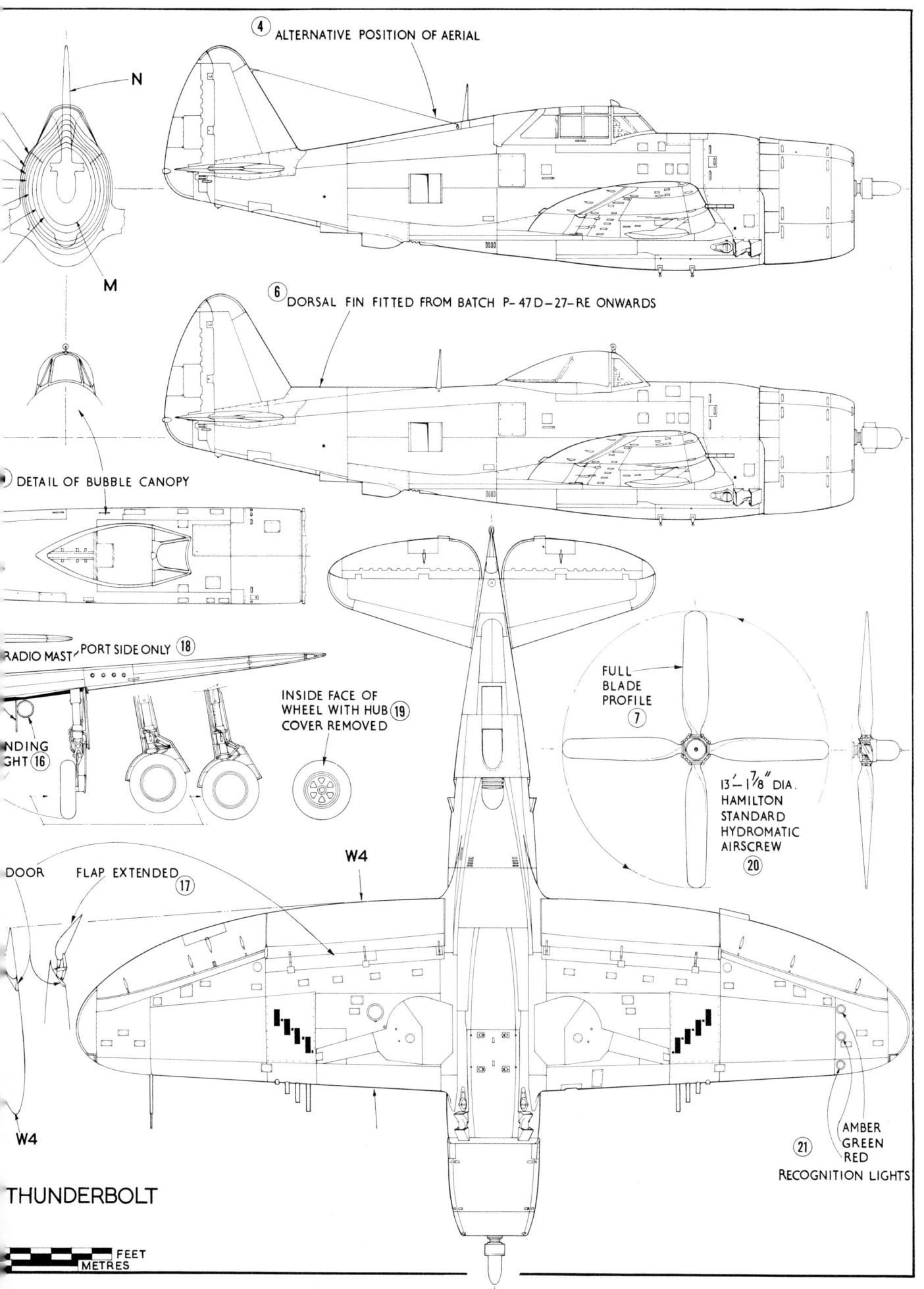

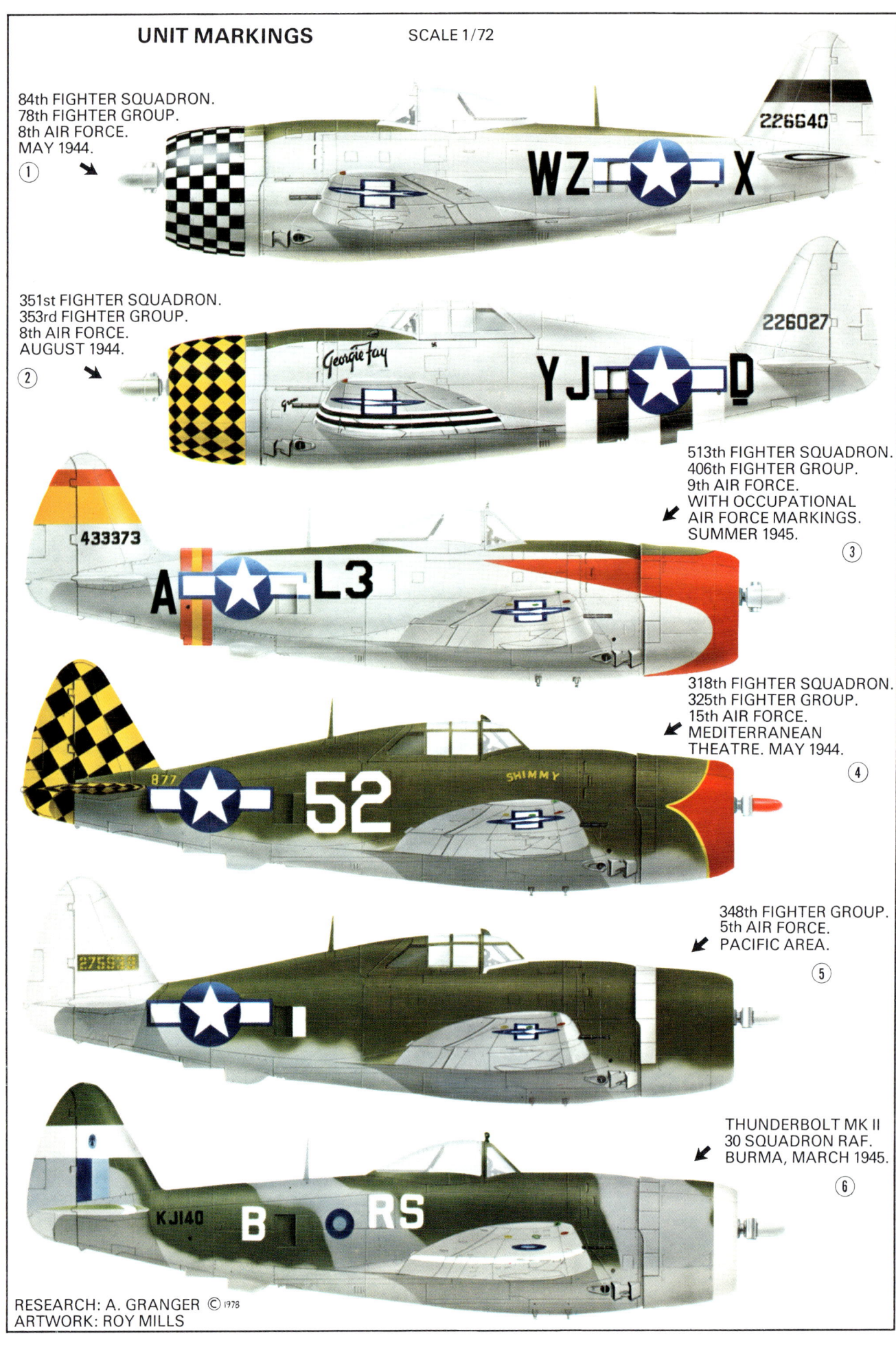

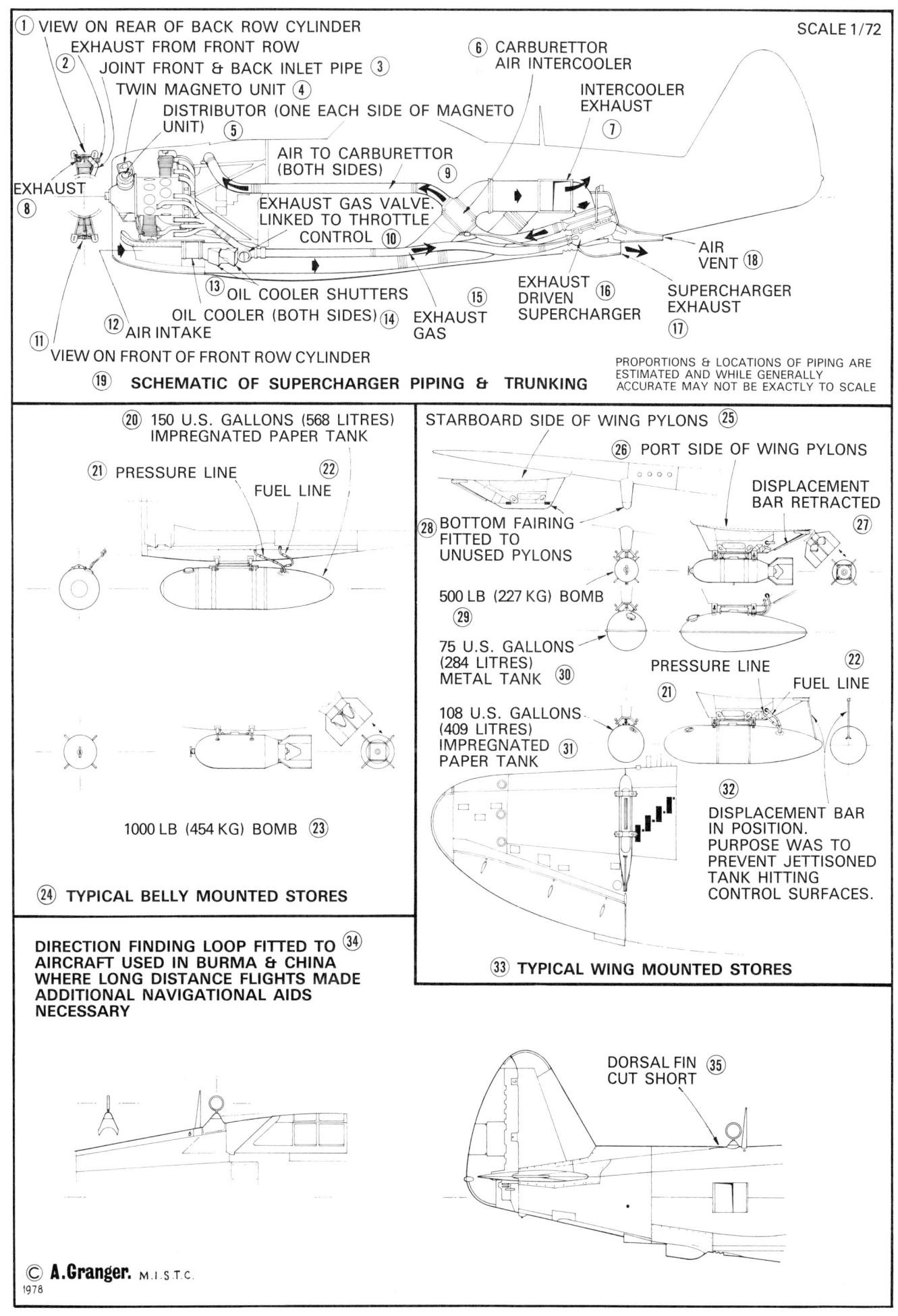

Figs. 10 & 11 *In the uppermost of these two additional views of 42-23278 the machine is seen to have a D/F aerial "egg" under its belly.* [Republic via R. A. Freeman]

By May 1944, the Fifteenth Air Force possessed seven fighter groups. Only two were equipped for a period of just under eight months with P-47Ds. The 325th Fighter Group, the celebrated "Checkertails", flew olive drab Thunderbolts carrying black and yellow group identification tail checks with the Fifteenth from November 1943 until April 1944. Their forward cowlings were red. Most Thunderbolts allocated to the 332nd FG in May and June 1944 came from the 325th. The conspicuous tail checks were overpainted olive drab or red. The unit's few new silver bubble-canopied P47Ds received red tails.

The 325th FG Thunderbolts flew 97 missions and 3,626 sorties mainly as escort for the Fifteenth's heavy bombers. They claimed 153 victories for the loss of 38 P-47Ds. Whilst flying the Thunderbolt six 325th FG pilots qualified as aces, and 66 pilots destroyed one or more enemy aircraft in the air. The unit received a Distinguished Unit Citation for mission 39. The group's Thunderbolts preceded the attacking bombers to the airfield at Villaorba where they destroyed 38 of the defending fighters and probably six more. The almost unmolested bombing blasted another 70 enemy planes on the ground. Only two Thunderbolts were lost.

Fig. 12 *The P-47K was radically modified from the last P-47D-5-RE, 42-8702, to have a bubble canopy, cut-down rear fuselage decking and repositioned radio equipment. It was completed 3 July 1943. Its star and bar markings had the current red surround.* [Republic via R. A. Freeman]

Fig. 13 *A bubble canopy P-47D-26-RA in natural metal finish and uncluttered by underwing pylons. Curtiss Electric propeller.* [Republic via R. A. Freeman]

Fig. 14 *A P-47D-25-RE carrying a 75 gallon (284 litre) belly tank and a 150 gallon (568 litre) tank on each wing pylon. Hamilton Standard propeller.* [Republic via R. A. Freeman]

Fig. 15 *A P-47D-25-RE, 42-26428, in flight clearly showing the improvement obtained by fitting a bubble canopy.* [Republic via R. A. Freeman]

Fig. 16 *One of 800 P-47D-30-REs built. This one, 44-20571, and coded D5-C, belonged to the 386th Fighter Squadron of the 365th Fighter Group of the Ninth Air Force.* [USAF via Harry Holmes]

Royal Air Force Thunderbolts served with South East Asia Command, although a few were evaluated in the United Kingdom. These included Thunderbolt I FL844 (USAAF No 42-25792) a "razorback" P-47D-22-RE, and three Mark IIs HD182, KJ298 and KJ299. The last RAF Thunderbolt II, KL887 (a P-47D-40-RA, USAAF No 44-90335), with electrically operated bubble canopy, dorsal fin fillet and underwing rocket clips, also came to this country, being finally scrapped at RAF Lichfield in 1946.

First RAF Thunderbolt squadron was No 146 which began to convert to Thunderbolt Is in May 1944. By September, with some Mark IIs as well, it was on armed reconnaissances and "Rhubarbs" on the Burma front. It was flying more than 1,000 sorties per month, first from Kumbhirgram and then Wangjing, liaising with the Army on "cab rank" duties between November 1944 and April 1945. Before the end of 1944, nine RAF squadrons were flying Thunderbolt Is and IIs, and by the end of the war, sixteen squadrons in South East Asia had re-equipped with the type. Three of these, Nos 81 (coded FL), 13 (NX) and 615 (KW) had converted from Spitfires; the rest, Nos 5 (OQ), 30 (RS), 34, 42 (AW); 60 (MU), 79 (NV), 113 (AD), 123 (XE), 134 (GQ), 135 (WK), 258 (ZT) and 261 (FJ) had previously slogged away with Hurricanes.

RAF Thunderbolts of both marks carried white recognition nose, chordwise wing and tail stripes on early green and grey camouflaged aircraft, and black or dark blue stripes on the later natural metal finish machines. Roundels and fin flashes were the standard light and dark blue common to all SEAC aircraft.

Fig. 17 *A razorback RAF Thunderbolt I, FL844 (USAAF No 42-25792) was a P-47D-22-RE retained in the United Kingdom for evaluation. It wore standard Fighter Command colours and is seen here in 1946 at RAF Lichfield awaiting scrapping.* [Author]

Fig. 18 *The very last RAF Thunderbolt II, KL887, a P-47D-40-RA (USAAF No 44-90335) in stained natural metal finish but with duck egg fuselage band, shallow dorsal fin fillet and underwing rocket clips, which was finally scrapped at RAF Lichfield in 1947.* [Author]

Only the evaluation Thunderbolts based in the United Kingdom and those that equipped No 73 OTU at Fayid in Egypt wore the duck-egg fuselage bands of Fighter Command machines.

Except for Alaska, the Thunderbolt operated on all active warfronts from early 1943. In the South West Pacific the Fifth Air Force was both the strategic and tactical air arm of General Douglas MacArthur's northward return from New Guinea back to the Philippines. Four of the Fifth's six fighter groups were at one time wholly or partly equipped with P-47Ds. The 348th Fighter Group introduced the P-47D Thunderbolt to Pacific operations in May 1943 and retained the type until January 1945 when the four squadrons of the group converted to P-51D Mustangs. On 11 October 1943, Colonel Neel E Kearby, the Group Commander in P-47D number 73, 42-8145 "Firey Ginger", personally shot down six aircraft from an escorted Japanese bomber formation. Kearby's flight of four, low on fuel, were returning home after a reconnaissance mission to Wewak. The action won him a Congressional Medal of honor, the first to be awarded to a USAAF fighter pilot. Colonel Kearby was killed in action on 5 March 1944. At the time he was the highest scoring P-47 ace in the SWPA with 22 victories.

After flying Airacobras and then Lightnings, all squadrons of the 35th Fighter Group converted to P-47Ds in November 1943 and then flew them on operations from early in 1944 until March 1945, when Mustangs replaced them. On the other hand only the 9th Fighter Squadron of the 49th Fighter Group had the P-47D and this from November 1943 to April 1944.

Lastly the 58th Fighter Group entered combat in February 1944 with "Razorback" P-47Ds. Its Thunderbolts shot down only 12 enemy aircraft, probably because the group's later role was almost exclusively a tactical and interdiction one. Sometime in 1943 the wing leading edges and tails were painted white on all 5th Air Force single-engined fighters. This was one of the first theatre markings to identify planes to ground troops.

When the Marianas were stormed by American forces in June 1944, P-47D Thunderbolts of the 318th Fighter Group of the Seventh Air Force flew off the decks of two escort carriers to land on Saipan the moment an airfield was available. From then on the group supported ground forces on Saipan, Tinian and Guam, strafed enemy airfields and patrolled US bases. Later from Ie Shima, the 318th FG, re-equipped from April 1945 with P-47Ns, joined other similarly equipped groups like the 508th FG of the Seventh Air Force and the 413rd, 414th and 507th Fighter Groups of the Twentieth Air Force to escort the new B-29s and also fly ground attack missions.

The Tenth Air Force and the Fourteenth Air Force together comprised the USAAF element in the China-Burma-India theatre. The 33rd Fighter Group moved to China in April 1944 to fly patrols and interdiction missions with their new Thunderbolts whilst their training was completed. Then in September 1944 the group returned to India and was reassigned to the Tenth Air Force. Their P-47Ds were employed mainly on strafing and dive bombing in Burma until VJ-Day.

The 80th Fighter Group was given the job of defending the Indian end of the transport route over the Himalayas, better known as the "Hump". Additionally its Thunderbolts hit Japanese airfields and safeguarded Allied airbases in the area. The group's Distinguished Unit Citation was awarded for a mission on 27 March 1944 against a large escorted Japanese bomber formation bent on destroying a large Assam oil refinery.

The 1st Air Commando Group which equipped with P-47Ds in May 1944 specialised in ground support in Burma and in bomber escort duties to Rangoon and elsewhere. Its mainly "razorback" Thunderbolts were distinctively marked with five diagonal stripes just aft of the cockpit and like many of the P-47s in the CBI theatre carried directional finding radio equipment with an external loop behind the pilot.

The 33rd Fighter Group was transferred to the US Tenth Air Force from the Fourteenth, whose other P-47D group was the 81st Fighter Group. It went to China in May 1944, but only became fully operational in January 1945.

Several models of the P-47 evolved and did not achieve production status, and quite a number of P-47Ds were used as development airframes. The radically different XP-47H was employed to test the Chrysler XIV-2220 sixteen cylinder liquid cooled inverted V engine, and not specifically to improve the performance of the Thunderbolt. The 2,500hp engine had several novel features and was designed to develop high power with minimum vibration. Two P-47D-15-RA airframes, 42-23297 and 42-23298, were set aside for conversion by Republic at Evansville under subcontract to Chrysler. The Thunderbolt's well-tried airframe was redesigned forward of the firewall and under the belly from the nose radiator to a point aft of the rear petrol tank. A General Electric CH5 turbo supercharger was mounted in the lower rear fuselage and a 13ft (3963mm) diameter four-bladed Curtiss Electric propeller with large conical spinner was fitted, this increasing the overall length of the XP-47H to 39ft 2in (11938mm). The wing guns were not installed: instead the space accommodated carbon dioxide for the flight tests. The first XP-47H flew on 26 July 1945 and made 27 flights before November. It achieved 490mph (789km/h) at 30,000ft (9144m).

The Curtiss-built P-47Gs and Republic-produced machines up to the P-47D-22-RE and P-47D-23-RA models all had framed sliding cockpit canopies and razorback rear fuselages. They suffered from a dangerous 20 degree blind spot to the rear. A blown glass "bubble" canopy giving all-round vision and a modified windscreen was fitted to and tested on the last

P-47D-5-RE (42-8702) which was redesignated XP-47K. The radio equipment was repositioned and later the XP-47K was the first Thunderbolt to have the long-range wing with integral fuel cells, similar in fact to that for the P-47N. The new canopy and modified rear decking radically changed the appearance of the standard service Thunderbolt. After successful tests the new style cockpit cover was introduced on production aircraft commencing with the P-47D-25-RE and P-47D-26-RA batches. Lateral instability caused by the cut-down rear fuselage and loss of keel area was corrected by a shallow dorsal fin added to the P-47D-27-RE and subsequent models.

The XP-47L was the last P-47D-20-RE (42-76614) on which was introduced a redesigned main fuel tank holding an extra 65 gallons (246 litres), six oxygen cylinders instead of four and other minor refinements. The changes made on the XP-47L were incorporated in the production Thunderbolts starting with the P-47D-25-REs.

The menace of the German V1 resulted in a rushed production model of the Thunderbolt. Three P-47D-27-RE airframes (42-27385, 42-27386 and 42-27388), were modified to take the Pratt and Whitney R-2800-57(C) engine and the larger CH5 turbo supercharger. The resulting war emergency power with water injection was 2,800hp at 32,500ft (9755m). Air brakes were fitted beneath the wings, and the YP-47Ms as they were designated, became the test prototypes for the 130 P-47Ms subsequently built. All the P-47M-1-REs were sent to the 56th Fighter Group in England where most were fitted with additional shallow dorsal fins.

The penultimate P-47D-27-RE (42-27387) out of 615 built was taken from the line and reworked as the XP-47N. Its fuselage was basically that of a P-47M. Its engine was a Pratt & Whitney R-2800-57 with CH5 turbo supercharger and regulator. Most important, however, it had the new blunt-tipped extended-span long-range wing eventually fitted to the single XP-47K. This was 18in (457mm) longer than the standard D wing and added 22sq ft (2.04sq m) to the wing area. It allowed space for four inter-connected wing root fuel cells which were self sealing and which gave the aircraft as much internal fuel again as the D model. Moreover, the undercarriage track was wider, the ailerons were bigger and the flaps larger. The model also had many detail improvements such as new tyres, wheels and brakes, an improved engine mount, redesigned engine cowl flaps and operating mechanism, increased oil tankage and double the previous water/alcohol capacity.

In combat, the P-47D Thunderbolt was equal to any fighter it met and superior to most when fitted with a paddle-bladed propeller and equipped with water injection. (The P-47D-11-RE was the first version to have water injection.) It was the heaviest single-engined single-seat fighter of the war. Yet it was able to out-perform enemy fighters at altitudes above 15,000ft (4572m), and with a superior rate of roll could carry out escape manoeuvres they were often unable to follow. The P-47 could both outdive the Bf109G and Fw190A and overhaul them in level flight. Similarly the weight which gave the Thunderbolt high speed in a dive could be used to augment the zoom climb. It was a steady gun platform with a terrific punch from its eight 0.50in (12.7mm) machine guns.

Engine failure dogged early P-47 operations for many weeks. The cause was attributed to blown cylinder heads resulting from manifold pressures which were too high and incorrect manipulation of turbo supercharger controls. Modification to the turbo superchargers and the fitting of inter-connected controls rectified the trouble, but not before a number of P-47s had been lost. The worst problem was radio interference. British VHF sets were fitted to all P-47s and changes made to magnetos, generators and sparking plugs.

The Thunderbolt's ability to absorb enormous damage and yet return home is shown by the narrow escape of top ace Robert S Johnson following his encounter with an Fw190 on 26 June 1943. He contrived to bring his battered P-47C-2-RE (41-6235 HV-P) back to Manston despite 21 jagged holes from 20mm cannon shells and more than 100 bullet holes. His P-47's torn cockpit cover was twisted and jammed, and would not slide back. Three 20mm cannon shells had burst against the armour plate behind Johnson's head and one had exploded in the cockpit next to his left hand, ripping away the flap handle. There were five cannon holes in the starboard wing and four in the port. Two cannon shells had blasted away the lower half of the rudder. With minimal forward vision through the oil spattered windscreen, yet unable to bail out, Johnson had to land the machine without flaps, brakes or hydraulics. His aircraft was a write-off. Johnson himself had bullet wounds to his nose and right thigh, burns on his forehead, shell splinters deeply embedded in both hands, and eyes badly swollen due to leaking hydraulic fluid.

Below 10,000ft (3048m) the Thunderbolt was very sluggish, especially when carrying full fuel and external stores. Manoeuvrability was poor, but as the altitude increased, handling qualities improved, until at high altitude the aircraft's performance was superlative. Performance was directly related to the efficiency of the turbo supercharger which gave a constant power output graph right up the altitude scale. For a fighter, the Thunderbolt had a big roomy comfortable cockpit regulated by a good temperature control.

As soon as Eighth Air Force headquarters in England realised there was a paramount need for fighter escorts to accompany Fortresses and Liberators on long daylight raids, VIIIth Fighter Command experimented with several types of jettisonable long-range tanks on its P-47s.

As early as 28 July 1943, P-47s from the 4th and 78th Fighter Groups, with belly tanks increasing their radius to 260 miles for the first time, gave escort cover to Eighth Air Force B-17s withdrawing from bombing north-west Germany. More than 100 P-47s, each with a 200gal (757 litre) unpressurised pressed paper belly tank, met the heavy bombers over the German-Dutch frontier as they returned from Kassel and Oschersleben (their deepest penetration to date). In the ensuing battle, the Thunderbolts claimed nine enemy fighters for the loss of one P-47. Then on 12 August 1943, the 56th Fighter Group joined in to fly its first droptank mission, using the same troublesome belly tanks. They were held to the P-47s by four-point suspension. There were all kinds of problems. The tanks were not good aerodynamically. Pilots found that they were unable to climb much above 20,000ft (6096m) without fuel starvation when drawing from the belly tanks. So the tanks were only partly filled and used to take the Thunderbolts

Fig. 19 *A P-47D-30-RE, one of the 446 P-47Ds that went to the Armée de l'Air. It has the normal French roundels (blue centres) with yellow outer surrounds and rudder flashes (blue foremost). Colouring is olive drab and neutral grey. Below the cockpit is the famous Sioux Indian head of the 1e Escadrille of GC II/5. It survives at the Chalais-Meudon museum.* [Harry Holmes]

up to that height and jettisoned. Fittings on the tanks were awkward and liable to break and patched tanks were common.

By the end of August 1943, the 75gal (284 litre) teardrop-shaped metal belly tanks in use were an improvement. They were combined with a pressurised system designed by an Air Technical Section team under Colonel Cass Hough which used the Thunderbolt's vacuum pump. They increased the aircraft's duration by about half an hour and its radius of action to 280 miles (451km). But these tanks did not allow fighters to escort the heavy bombers into the heart of Germany where fighter protection would become imperative.

It was not until the first 108gal (409 litre) metal drop tanks were delivered in early September and similar capacity British-made paper tanks were available that Thunderbolts could escort the B-17s right into German airspace. On 27 September whilst escorting the heavies to Emden, Thunderbolts carrying 108gal tanks flew 400 miles mainly over water and inflicted a 21 to 1 defeat over defending Luftwaffe fighters.

Each British-made 108gal paper tank was formed in three sections. The centre section containing two plywood baffle bulkheads, metal filler cap, vent pipe and fuel draw connection, was straight wound in paper until a wall thickness of $^{5}/_{16}$in (8mm) was achieved. Projecting internal reinforcing rings at each end of the centre section facilitated the joining of separately moulded nose and tail sections. The adhesive was urea formaldehyde.

Introduced at the beginning of 1944 were improved P-47D Thunderbolts with fuselage and wing strongpoints for bombs or fuel tanks. Combat radius with a single 150gal (568 litre) belly tank was increased to 425 miles (684km), and with two 150gal wing tanks to 475 miles (765km). Machines from P-47D-25-REs onwards had increased main tankage (370gal or 1400 litres) and provision for a total fuel load of up to 780gal (2954 litres). Combat radius was beyond Berlin and Thunderbolts flew over the city for the first time on escort duty in March 1944.

SPECIFICATION

Powerplant: One Pratt & Whitney R2800 or -59 eighteen-cylinder radial air-cooled engine of 2,300hp or 2,535hp.
Dimensions: Span 40ft 9$^{5}/_{16}$in (12429mm); length 36ft 1$^{3}/_{16}$in (11003mm).
Weights: Empty 10,000lb (4536kg), normal loaded 14,000lb (6350kg), maximum loaded 17,500lb (7734kg).
Performance: Max speed 353mph (568km/h) at 5,000ft (1524m), 406mph (653km/h) at 20,000ft (6096m), 433mph (697km/h) at 30,000ft (9144m). Max rate of climb 2,750ft/min (838m/min) at 5,000ft (1524m), 2,140ft/min (642m/min) at 20,000ft (6096m). Time to altitude 4.3min to 10,000ft (3048m), 11.0min to 20,000ft (6096m). Service ceiling 42,000ft (12801m). Range 640 miles (1030km) at 278mph (447.4km/h) at 25,000ft (7620m) (internal fuel only), 925 miles (1488.6km) at 231mph (371.8 km/h) at 10,000ft (3048m) (max external fuel).
Armament: Eight .50in (12.7mm) Browning machine guns with 267 or 425rpg plus three 500lb (227kg) or two 1,000lb (454kg) bombs on underwing pylons, or underwing rockets.

Fig. 20 *A P-47D-40-RA of the USAF photographed post WW2. It had the large "buzz" number FF-431 on the fuselage sides, red nose, wing tips, fin tip and diagonal stripes. The top of the fuselage, including the shallow dorsal fin, was painted olive drab.* [Harry Holmes]